4th International Conference on Measurement Instrumentation and Electronics (ICMIE 2020)

MATEC Web of Conferences Volume 327 (2020)

Chengdu, China
18 - 20 September 2020

Editor:

Ab Wafi Bin Ab Aziz

ISBN: 978-1-7138-2100-7

Printed from e-media with permission by:

Curran Associates, Inc.
57 Morehouse Lane
Red Hook, NY 12571

Some format issues inherent in the e-media version may also appear in this print version.

For additional information, please contact EDP Sciences – Web of Conferences at the address below.

EDP Sciences – Web of Conferences
17, Avenue du Hoggar
Parc d'Activité de Courtabœuf
BP 112
F-91944 Les Ulis Cedex A
France

Phone: +33 (0) 1 69 18 75 75
Fax: +33 (0) 1 69 28 84 91

contact-edps@webofconferences.org

Additional copies of this publication are available from:

Curran Associates, Inc.
57 Morehouse Lane
Red Hook, NY 12571 USA
Phone: 845-758-0400
Fax: 845-758-2633
Email: curran@proceedings.com
Web: www.proceedings.com

TABLE OF CONTENTS

PREFACE

This volume contains papers presented at 2020 4th International Conference on Measurement Instrumentation and Electronics (ICMIE 2020), which was held during September 18-20, 2020 in Chengdu, China.

ICMIE 2020 provides a scientific platform for both local and international scientists, engineers and technologists who work in all aspects of measurement instrumentation and electronics. The goal of ICMIE is to establish an international forum for engineers and scientists to present their ideas and experiences in the fields of measurement instrumentation and electronics like we always do.

The volume includes 18 selected papers which were submitted to the conference from universities, research institutes and industries. Each contributed paper has been peer-reviewed by reviewers who were collected organizing and technical committee members as well as other experts in the field from different countries. The proceedings tend to present to the readers the newest researches result and findings in the field of measurement instrumentation and electronics.

The chairpersons of different sessions played important role in conducting the proceedings of the session in a timely and efficient manner, and on behalf of the conference committee, we express sincere appreciation for their involvement. The reviewers of manuscripts have also been very helpful in efficiently reviewing the manuscripts, providing valuable comments well within the time allotted to them. We express our sincere and grateful thanks to all the sponsors and participants for their valuable contributions and support to ICMIE 2020.

The expectation is that this conference is a significant contribution to this area and that the organization deserves your admiration.

ICMIE 2020 Conference Committee

edp sciences

Web of Conferences

Statement of Peer review

In submitting conference proceedings to *Web of Conferences*, the editors of the proceedings certify to the Publisher that

1. They adhere to its **Policy on Publishing Integrity** in order to safeguard good scientific practice in publishing.
2. All articles have been subjected to peer review administered by the proceedings editors.
3. Reviews have been conducted by expert referees, who have been requested to provide unbiased and constructive comments aimed, whenever possible, at improving the work.
4. Proceedings editors have taken all reasonable steps to ensure the quality of the materials they publish and their decision to accept or reject a paper for publication has been based only on the merits of the work and the relevance to the journal.

Title, date and place of the conference

2020 4th International Conference on Measurement Instrumentation and Electronics
Chengdu, China
September 18-20, 2020

Proceedings editor(s): Dr. AB WAFI BIN AB AZIZ

Date and editor's signature

26 October 2020

17, avenue du Hoggar - PA de Courtabœuf – BP 112 - 91944 Les Ulis Cedex A (France)
Tél. : 33 (0)1 69 18 75 75 - Fax : 33(0)1 69 07 45 17 – www.edpsciences.org

MATEC Web of Conferences 327, 01001 (2020) https://doi.org/10.1051/matecconf/202032701001
ICMIE 2020

Monte Carlo Method to Uncertainty Evaluation of The Relative Dielectric Constant and Loss Tangent Measured by Split-Cavity Resonator Technique

Fei Zhao[1,*], *Jing* Pei[1], *Qi* Tang[2], *Chengxiang* Wang[3] and *Jinsong* Kan[1]

[1]Metrology and Testing Center, China Electronics Standardization Institute, Beijing 100176, China
[2]School of Electronic Engineering, Xi'an University of Posts & Telecommunications, Xi'an 710121, China
[3]Shandong Qinghua Tongfang Luying Electronic Co., Ltd., Yinan County Town 276300, Shandong, China

Abstract. According to the mathematical model of the split-cylinder resonator method, it is impossible to establish a simple equation between the intermediate variables and the relative dielectric constant or the loss tangent. Instead, the measured relative dielectric constant and the loss tangent must be calculated numerically by iterative programming. When evaluating uncertainty according to the Guide to the Expression of Uncertainty in Measurement (GUM), it is impossible to obtain the sensitivity coefficient of each intermediate. Therefore, the reliability of uncertainty evaluation is greatly reduced. ISO/TS15530 states that the most effective method for estimating uncertainty is computer simulation, and more specifically, a Monte Carlo simulation. In this study, Three typical samples with different relative dielectric constant and loss tangent were measured, respectively. The probability density function of each intermediate variable was given, according to which the Monte Carlo simulations were performed by MATLAB programming. Moreover, some of the simulation results are compared with those of the previous literature, and the results show that the measurement uncertainty evaluated by the GUM method was slightly larger, indicating that it was a little bit conservative to assume that each intermediate variable is completely uncorrelated in the GUM method.

1 Introduction

For measuring the relative dielectric constant and loss tangent of low-loss dielectric substrates non-destructively, The split-cylinder resonator technique method was originally proposed by G. Kent [1]. Janezic and Baker-Jarvis [2] made a full-wave analysis of the split cylinder resonator and outlined a model where the fringing fields are rigorously accounted for and the resonance condition is derived. Moreover, by introducing a proper boundary, they enclosed the entire split-cylinder resonator fixture, thereby enabling them to derive a valid resonance condition using the mode-matching method without introducing a systematic error in the calculation of the relative dielectric constant and loss tangent of the samples [3,4]. Fang and Linton et al. [5,6] also suggested a developed apparatus and made a systematic analysis based on a rigorous mode match model.

According to the Guide to the Expression of Uncertainty in Measurement (GUM), the uncertainty of measurements was evaluated by Janezic [3], and similar uncertainty results were also obtained on the basis of our own split-cylinder resonator [7]. However, there are obvious deficiencies in the mathematical model of measurement. Both the relative dielectric constant (ε_r) and loss tangent ($\tan\delta$) of the sample are assumed to be abstract functions of several intermediate variables. Therefore, the sensitivity coefficients of all intermediate variables cannot be directly calculated. They are all assumed to be non-related instead.

ISO/TS15530 states that the most effective method for estimating uncertainty is computer simulation, and more specifically, a Monte Carlo simulation. To the best of the knowledge of the authors, for measurement or calibration of the ε_r and $\tan\delta$ of the samples measured by a split-cylinder resonator method, there are few reports on studies that have evaluated the measurement uncertainties by Monte Carlo method (MCM) and then verified the equivalence of the measurement uncertainties by GUM and MCM. Therefore, In this study, detailed uncertainty evaluation procedure has been given by applying Monte Carlo simulation for evaluating the uncertainties of the ε_r and $\tan\delta$ of the samples measured by a split-cylinder resonator method. The Monte Carlo simulation results of three typical samples with different the ε_r and $\tan\delta$ are presented. Some of the results were compared with those of the previous literature.

* Corresponding author: cesizf@163.com

2 Methodology

2.1 Monte Carlo method of estimating measurement uncertainty

The Monte Carlo method (MCM) is a tool that uses the computational capacity of current computers to simulate a high account of pseudo random numbers. In this way, it allows to simulate complex system from a probabilistic point of view. The MCM uses a large number of samples, with different probability density functions, to obtain the final uncertainty distribution. Figure 1 illustrates the propagation of distributions. The model input quantities are $X = (X_1, X_2, ..., X_N)^T$. The probability density functions (PDF) $g_{Xi}(\xi_i)$ may be different from each other. The PDF $G_Y(\eta)$ for the value of the output quantity Y is indicated as being asymmetric, as can arise for nonlinear models. Therefore, a large set of model values obtained can be applied to provide an approximation to the distribution of possible values for the output quantity.

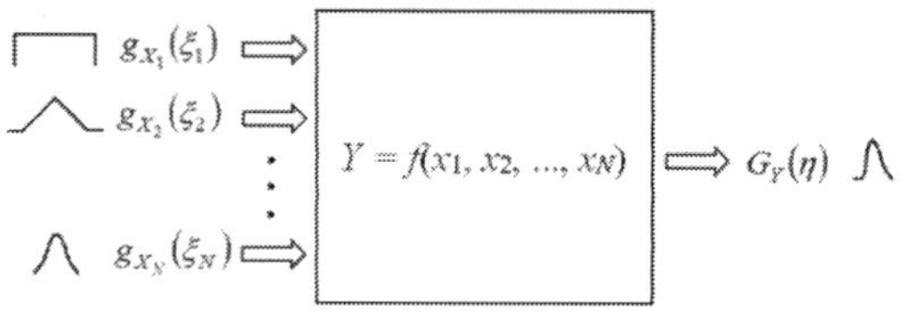

Fig. 1. Illustration of the propagation of distribution based on MCM.

An implementation procedure of the propagation of distributions based on MCM can be stated as a step-by-step procedure:

1) The model $Y = f(X_i)$ shall be established at the formulation stage of uncertainty evaluation;
2) The probability density functions for the input quantities X_i, $g_{Xi}(\xi_i)$ shall be assigned;
3) The number of M of Monte Carlo trails to be made shall be selected;
4) The required coverage probability p (e.g. 95%, or 0.95) shall be determined;
5) M samples of the (set of N) input quantities shall be generated;
6) A model giving the corresponding output quantity value shall be evaluated for each sample;
7) The values of the output quantity shall be sorted into non-decreasing order, using the sorted values to approximate the distribution function for the output quantity value;
8) The estimate of the output quantity value and the associated standard uncertainty from the distribution function shall be formed;
9) The shorted p coverage interval for the output quantity value from this distribution function shall be formed.

For more details about the MCM of estimating measurement uncertainty, one can refer to Ref [8-10] or other related literature.

2.2 Split-cylinder Resonator techniques

The principal drawing of the split-cylinder resonator is shown in Figure 2. A planar dielectric sample is placed in the variable-length gap between the two shorted cylindrical waveguide sections. In order to excite the TE_{0np} family of resonant modes, coupling loops in both waveguide sections are connected to the input ports of a network analyzer via coaxial transmission lines. From measurements of the resonant frequency and quality factor of the TE_{0np} resonant modes, the ε_r and $\tan\delta$ of the sample can be determined. In the model for calculating the compex dielectric constant, Janezic et al. [2-4] assumed that a perfect conductor exists at some sufficiently large $\rho = b$. By introducing this boundary, they enclose the entire split-cylinder resonator fixture, thereby enabling them to derive a valid resonance condition using the mode-matching method without introducing a systematic error in the calculation of the complex dielectric constant of the sample. The derivation of the ε_r and $\tan\delta$ was given in literature [2-4]. In this work, the numerical calculation of the above derivation is realized by MATLAB programming [11, 12], by which the uncertainty evaluation based on MCM is also achieved.

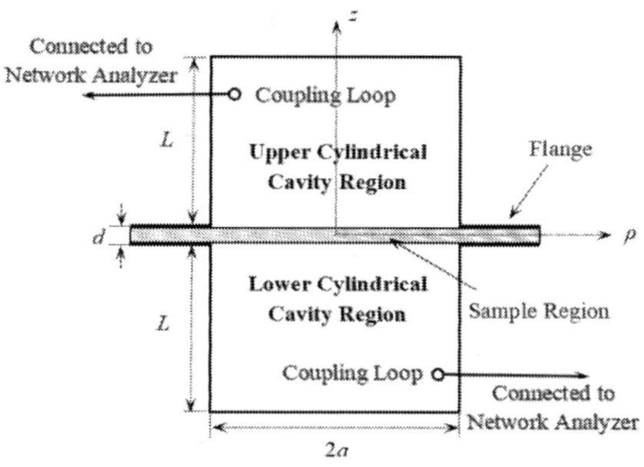

Fig. 2. Cross-sectional diagram of a split-cylinder resonator.

Three typical samples with nominal relative dielectric constant ranging from 3.8 to 77 and loss tangent ranging from 8×10^{-5} to 4×10^{-4} have been measured using our own split-cylinder resonator fixture, of which the nominal resonant frequency of TE_{011} mode is 10 GHz. An E8363B network analyzer (Agilent Technologies, CA) was used in the measurement.

2.3 Mathematical model of the measurement

According to the mathematical model of the split-cylinder resonator technique, the ε_r and $\tan\delta$ should be programmed for iterative numerical calculations. Therefore it is impossible to establish a simple equation between the intermediate variables and the ε_r and $\tan\delta$. However, these two parameters can be assumed as a function f of several intermediate variables:

$$\varepsilon_r = f(f_1, a, L, d) \tag{1}$$

$$\tan\delta = f(f_1, Q_1, Q_0, a, L, d) \tag{2}$$

where:

f_1 — the resonant frequency of the TE_{011} mode of an cavity with a sample

Q_1 — the quality factor of the TE_{011} mode of an cavity with a sample

Q_0 — the quality factor of the TE_{011} mode of an cavity without a sample

a — cylindrical cavity radius

L — cylindrical cavity length

d — sample thickness

In advance of dielectric constant measurements, the cavity parameters such as cylindrical cavity radius and length are determined from the measurements for the TE_{011} and TE_{012} resonance modes of an empty cavity without a sample. The cylindrical cavity radius and length are determined from two measured resonant frequencies, f_{01} for the TE_{011} mode and f_{02} for the TE_{012} mode, by using the following equations [13]:

$$a = \frac{cj'_{01}}{2\pi}\sqrt{\frac{3}{4f_{01}^2 - f_{02}^2}} \quad (3)$$

$$L = \frac{c}{4}\sqrt{\frac{3}{f_{02}^2 - f_{01}^2}} \quad (4)$$

where:

c — the velocity of light in a vacuum, $c = 2.9979\times10^8$ m/s

j'_{01} — the first zero of the Bessel function of the first kind J_1, $j'_{01} = 3.83173$

f_{01} — the resonant frequency of the TE_{011} mode

f_{02} — the resonant frequency of the TE_{012} mode

3 Results and Discussion

3.1. The input quantity and its PDF

Intermediate variables related to the cavity parameters are used in the ε_r and tanδ calculation. Table 1 lists each intermediate variable and its assigned Probability density functions (PDF).

Table 1. Input values and PDF for the cavity parameters.

Quantity X_i	Probability density functions	Silica
f_{01}	Rectangular distribution, R(a,b)	R(10017.64 MHz, 10017.70 MHz)
f_{02}	Rectangular distribution, R(a,b)	R(11234.74 MHz, 11234.80 MHz)

Three typical samples with different nominal relative dielectric constant and loss tangent have been measured and the intermediate variables, together with assigned PDF are listed in Table 2, Table 3 and Table 4, respectively.

Table 2. Input values and PDF for the sample Silica evaluation.

Quantity X_i	Probability density functions	Silica
f_1	Rectangular distribution, R(a,b)	R(9339.77 MHz, 9339.83 MHz)
d	Normal distribution, N(μ,σ)	N(1.007 mm, 0.004 mm)
Q_1	Rectangular distribution, R(a,b)	R(12368, 13133)
Q_0	Rectangular distribution, R(a,b)	R(25608, 27192)

Table 3. Input values and PDF for the sample K20 evaluation.

Quantity X_i	Probability density functions	K20
f_1	Rectangular distribution, R(a,b)	R(6301.56 MHz, 6301.62 MHz)
d	Normal distribution, N(μ,σ)	N(0.976 mm, 0.004 mm)
Q_1	Rectangular distribution, R(a,b)	R(9021, 9579)
Q_0	Rectangular distribution, R(a,b)	R(25608, 27192)

Table 4. Input values and PDF for the sample K77 evaluation.

Quantity X_i	Probability density functions	K77
f_1	Rectangular distribution, R(a,b)	R(3445.53 MHz, 3445.59 MHz)
d	Normal distribution, N(μ,σ)	N(1.003 mm, 0.004 mm)
Q_1	Rectangular distribution, R(a,b)	R(2134, 2266)
Q_0	Rectangular distribution, R(a,b)	R(25608, 27192)

3.2. The number of Monte Carlo trials

The number of Monte Carlo trials was set to 1000000. Since the ε_r and tanδ obtained by the split-cylinder resonator method are solved by numerical iterative method, it takes about 3 to 5 seconds to calculate the results of each time on a workstation (Lenovo D20; CPU: Intel Xeon E5649 2.53GHz, 2 processors with 12 GB RAM). Therefore, it would have taken about 1000 ~ 1400 hrs to finish totally 1000000 times simulations. In order to shorten the simulation time, the trails on the workstation were divided into 10 independent parallel calculations. As a result, the number of each trials dropped to 100000. Even so, it still takes about 200 ~ 220 hrs to complete all the calculations for each samples.

3.3. Uncertainty of relative dielectric constant

Figure 3 shows the Monte Carlo simulation's resulting histogram with corresponding normal fit and the shortest 95% coverage interval for the ε_r of the sample Silica, K20 and K77, respectively. These results reveal that the probability density function of calibrated ε_r of all samples with different ε_r shows a normal distribution.

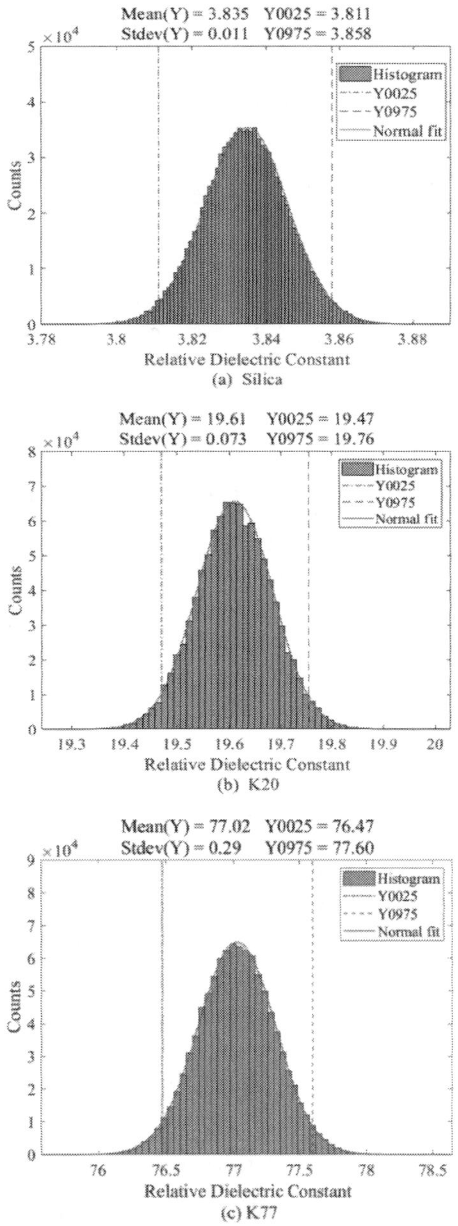

Fig. 3. Probability density function for relative dielectric constant calibration of the sample Silica, K20 and K77.

The calibrated relative dielectric constant for the sample Silica is 3.835 and the corresponding uncertainty is 0.024, that is, the relative uncertainty is 0.63% (k = 2). The uncertainty is much smaller than that of the GUM method[3,7]. The calibrated value for the sample K20 is 19.61 and the corresponding uncertainty is 0.15, that is, the relative uncertainty is 0.77% (k = 2). The uncertainty is also smaller than that of the GUM method obtained in our previous study [7]. These results may indicate that it was slightly conservative to assume the sensitivity coefficient of each intermediate variable is completely uncorrelated in both previous GUM method [3,7]. The calibrated value for the sample K77 is 77.02 and the corresponding uncertainty is 0.57, that is, the relative uncertainty is 0.74% (k = 2).

3.4. Uncertainty of loss tangent

Figure 4 illustrates the Monte Carlo simulation's resulting histogram and the shortest 95% coverage interval for the tanδ of the sample Silica, K20 and K77, respectively.

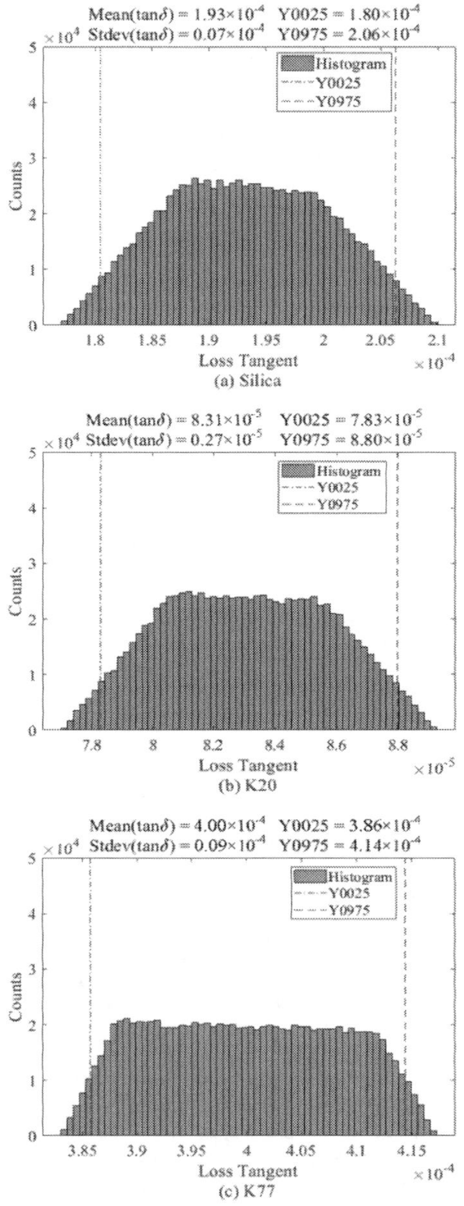

Fig. 4. Probability density function for loss tangent calibration of the sample Silica, K20 and K77.

These results reveal that the probability density function of calibrated loss tangent of all samples with different loss tangent shows a trapezoidal distribution. The calibrated loss tangent for the sample Silica is 1.93×10^{-4} and the corresponding uncertainty is 0.13×10^{-4} (k = 2). The uncertainty is on the same order of magnitude as that evaluated by GUM given in Ref [3]. The calibrated loss tangent for the sample K20 is 8.3×10^{-5} and the corresponding uncertainty is 0.49×10^{-5} (k = 2). The calibrated loss tangent for the sample K77 is 4.00×10^{-4} and the corresponding uncertainty is 0.14×10^{-4} (k = 2).

4 Conclusions

Measurement uncertainty is used to rationally characterize the dispersion of a measurement value and parameters related to measurement results, helping ensure the accuracy and reliability of measurement results and the accuracy of judgment regarding whether products conform to requirements, assisting in the acquisition of laboratory accreditation. In response to actual requirements, this paper demonstrated estimations of the uncertainty of the ε_r and tanδ measurements by split-cylinder resonator method using Monte Carlo simulations run by MATLAB programming. In addition, some of the simulation results were compared with those of the previous literature, which showed that the measurement uncertainty evaluated by GUM method may overestimate some uncertainty components, resulting in slightly larger evaluation results.

Since the current Monte Carlo simulation process is too time-consuming, the adaptive Monte Carlo method should be further studied. If we can prove that these two methods are equivalent, from the time-consuming point of view, the adaptive Monte Carlo method may be more practical for the uncertainty evaluation method of the ε_r and tanδ of dielectric samples.

References

1. G. Kent. *An evanescent-mode tester for ceramic dielectric substrates*. IEEE T. Microw. Tech. **36**, 1451 (1988)
2. M.D. Janezic, J. Baker-Jarvis. *Full-wave analysis of a split-cylinder resonator for nondestructive permittivity measurements*. IEEE T. Microw. Tech. **47**, 2014 (1999)
3. M.D. Janezic. *Nondestructive relative dielectric constant and loss tangent measurement using a split-cylinder resonator* (Ph. D. Thesis, University of Colorado at Boulder, 2003)
4. M.D. Janezic, E.F. Kuester, J. Baker-Jarvis. *Broadband complex permittivity measurements of dielectric substrates using a split-cylinder resonator*. IEEE MTT-S Int. Microw. Symposium Digest, 1817 (2004)
5. X.Y. Fang, D. Linton, C. Walker, B. Collins. *Non-destructive characterization for dielectric loss of low permittivity substrate materials*. Meas. Sci. Technol. **15**, 747 (2004)
6. X.Y. Fang, D. Linton, C. Walker, B. Collins. *A tunable split resonator method for nondestructive permittivity characterization*. IEEE T. Instrum. Meas. **53**, 1473 (2004)
7. F. Zhao, P Xu, C.T. Sha, W.F. Wang, J.S. Kan. *A nondestructive method for measurements of complex permittivity of microwave dielectric substrate materials*. Proc. 3rd Int. Conf. Mater. Mech. Manuf. Eng. 968 (2015)
8. JCGM 104:2009. *Evaluation of measurement data —An introduction to the "Guide to the expression of uncertainty in measurement " and related documents*. (Joint Committee for Guides in Metrology, Sèvres, 2009)
9. JCGM 101:2008. *Evaluation of measurement data — Supplement 1 to the "Guide to the expression of uncertainty in measurement " — Propagation of distributions using a Monte Carlo method*. (Joint Committee for Guides in Metrology, Sèvres, 2008)
10. JCGM 102:2011. *Evaluation of measurement data — Supplement 2 to the "Guide to the expression of uncertainty in measurement " — Extension to any number of output quantities*. (Joint Committee for Guides in Metrology, Sèvres, 2011)
11. C.P. Robert, G. Casella, *Monte Carlo Statistical Methods* (Springer-Verlag, New York, 1999)
12. M. Fernández, J.M. Calderón, P.M. Bravo Díez. *Implementation in MATLAB of the adaptive Monte Carlo method for the evaluation of measurement uncertainties*. Accred. Qual. Assur. **14**, 95 (2009)
13. IEC 62562:2010. *Cavity resonator method to measure the complex permittivity of low-loss dielectric plates*. (European Committee for Electrotechnical Standardization, Brussels, 2011)

Study on the Technology of Ultra-high Sensitive Wide-band Magnetic-feedback Inductive Magnetic Sensor

Yong Liu*, *Wenbin* Li, *Jinrong* Zhou, *Rui* Pan, *Huan* Zheng, *Bing* Xiang, and *Rui* Xu

LF Electromagnetic Communication Laboratory, Wuhan Maritime Communication Research Institute, 430205 Wuhan, China

Abstract. It analyses the composition and principle of high-sensitive wide-band magnetic-feedback inductive magnetic sensor to fulfil the demand of high-sensitive wide-band magnetic sensor in geological exploration. It studies main factors to the performance of wide-band magnetic sensor, such as turns of coils, core material features, and amplifier noise, specifies section-wise coil winding, the type and dimension of core material, and designs low-noise high-impedance LF chopping amplifier channel and composite amplifier with HF amplifier channel. The noise of magnetic sensor at 1 Hz is better than 10-4nT/Hz1/2, at 100Hz-1kHz band the noise floor is close to SQUID which can reach 10-6nT/Hz1/2. The magnetic sensor works at wide frequency band (0.0001Hz-10kHz) and ultra-low noise, which can meet the requirements of both AMT and CSAMT.

1 Introduction

High-sensitive wide-band inductive magnetic sensor is the most widely-applied magnetic receiver in electromagnetic prospecting, mainly used in the exploration of metal mineral, petroleum resources, and in-depth geophysics[1-3].Electromagnetic prospecting is divided into audio frequency magnetelluric (AMT) and Controlled source audio frequency magnetelluric (CSAMT), with the former requiring high-sensitive magnetic sensor, and the latter wide frequency-band, meanwhile the sensor shall fulfil field requirements like small size, light weight, and low power consumption[2]. Based on actual field environment, the paper studies how to realize wide frequency-band and low-noise of inductive magnetic sensor from theoretical and field design perspective, which can fulfil the requirements of both AMT and CSAMT exploration.

2 Composition and principle

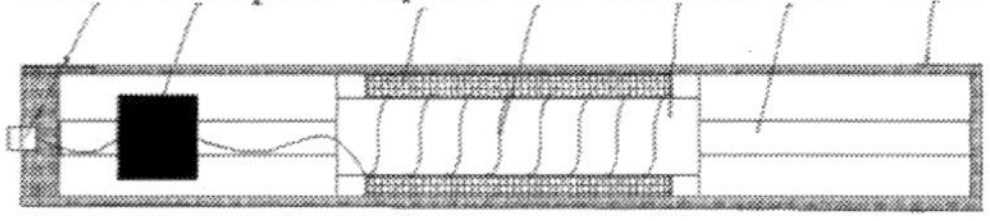

Fig. 1. Schematic of magnetic sensor composition.

Magnetic-feedback inductive magnetic sensor is mainly composed of core, coil (inductive coil and feedback coil included), pre-amplifier, coil frame, interface and insulating shell, the schematic of which is shown as Figure 1. The sensor is capsulated in the insulated cylinder with water-tightness and vibration-proof features, suitable for field environment and preventing water and mechanical shock force.

The basic principle of magnetic sensor is Faraday's law of electromagnetic induction, and inducted electromotive force U_i will be generated in case that cylinder coil, with N being the number of turns and S being sectional area, is placed in the magnetic field $B(t)$ varied with time, which is shown as follows:

$$U_i = -NS\frac{dB(t)}{d(t)} = -j\omega NS\mu_r B_0 \qquad (1)$$

In the formula N is the number of turns, S sectional area, μ_r relative permeability of core, and B_0 being magnetic field component in the direction of core.

It is known from formula (1) that the increase of N, S, μ_r will directly enhance field induction capacity of coil [4], meanwhile reducing the local noise can enhance SNR of output signal, and raise magnetic induction capacity indirectly.

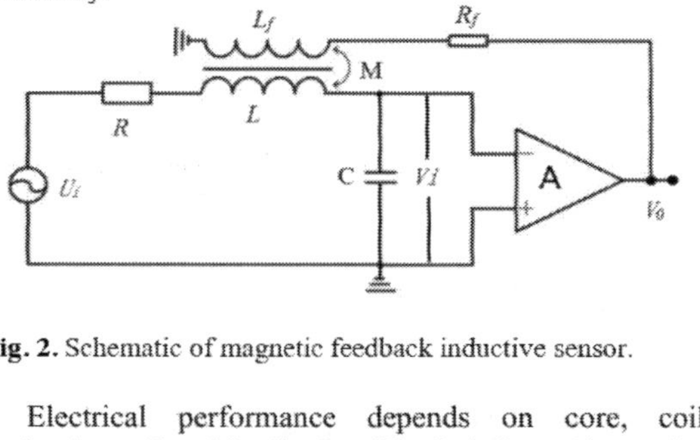

Fig. 2. Schematic of magnetic feedback inductive sensor.

Electrical performance depends on core, coil (inductive coil and feedback coil included), and low noise pre-amplifier. As shown in Figure 2, R, L, C respectively are resistance, inductance, and distributive capacitance of inductive coil. Through the amplification of amplifier A,

* Corresponding author: 13886062187@139.com

inducted output voltage of L is entered into L_f via R_f, in which a feedback magnetic field is formulated when there is current passing by, therefore minus feedback is established against the tested magnetic field, and minus compensation is carried out over the output signal at frequency band adjacent of resonance point, so that amplitude curve is smooth and even for output signal at these frequencies, in an effort to enhance bandwidth of magnetic sensor [5]. Generally, feedback is adopted to spread the frequencies upward to 10kHz above, and chopping amplifier to spread the frequencies downward to 0.0001Hz. Magnetic feedback can reduce the effect of temperature, humidity, and other factors on the sensor, enhancing consistency and stability. Usually feedback coil serves as nominal coil for calibrating magnetic sensor in the field.

Generally a layer of feedback coil is winded on inducted coil, with its inductance being L_f, resistance being R_f , and mutual inductance being M via formula $M = Q\sqrt{L * L_f}$; close coupling between inducted coil and feedback coil is in place with its factor Q=1, hence $M = \sqrt{L * L_f}$. And transfer function $H(\omega)$ is shown as follows suppose gain of amplifying circuit being A, and output being V_0:

$$H(\omega) = \frac{V_o}{B_0} = \frac{j\omega\mu_r NSA}{1-\omega^2 LC + j\omega(RC + \frac{AM}{R_f})} \tag{2}$$

It is found that LF stop frequency of amplitude-frequency characteristic is $f_l = \frac{R_f}{2\pi AM}$, and HF stop frequency being $f_h = \frac{AM}{2\pi R_f LC}$ based on transfer function. Therefore under the situation that feedback resistance is certain, i.e. feedback depth being constant, and that amplifying product A of amplifying circuit is increased, the more feedback factor α is, LF stop frequency will be decreased, and HF stop frequency will be increased, which is conducive for spreading the width of frequency band.

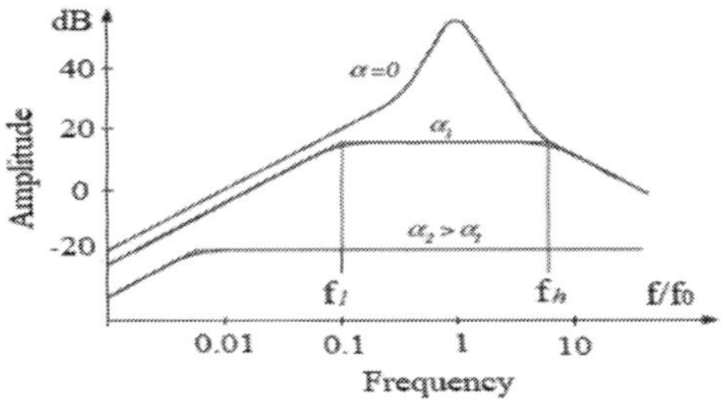

Fig. 3. Amplitude-frequency curve for magnetic sensor with different feedback factor α.

Figure 3 shows amplitude-frequency curve for magnetic sensor with different feedback factor α, from which the more negative feedback factor is, the deeper negative feedback will be, and the more frequency range will be in case of amplitude being flat, which is convenient for processing testing data. Magnetic feedback has the same effect on field signal and noise of amplifier input, not changing output SNR.

3 Raising magnetic induction capacity

3.1 Raising the number of coil turns via section-wise winding

Sensitivity can be enhanced via raising the number of coil turns, however it also results in the increase of coil inductance, resistance, distributive capacitance, reduction of resonance frequencies, and increase of resistance noise. Too much inductance may hardly match with amplifier, and amplifier noise current breeds more high frequency noise.

Section-wise winding without regulations can reduce inductance and distributive capacitance, as shown in Figure 4 the schematic of equivalence between inductance and distributive capacitance after sectioning. In case of total number of turns being certain, as the number of sections is increased, mutual induction and distributive capacitance will be reduced between sections, owing to which total induction and distributive capacitance can be reduced. In case of total number of turns being 11000, simulation is carried out for inductance when X=2-20, the result is shown as Figure 5, from which the total induction can be decreased to 1/7 of the original value in case of the number of sections from 2 to 20 [5].

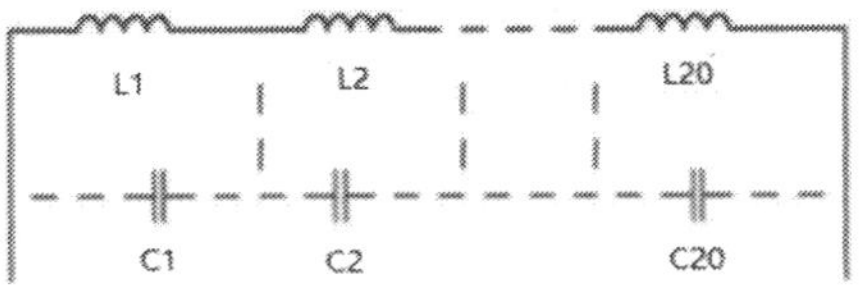

Fig. 4. Schematic of equivalence between inductance and distributive capacitance after sectioning.

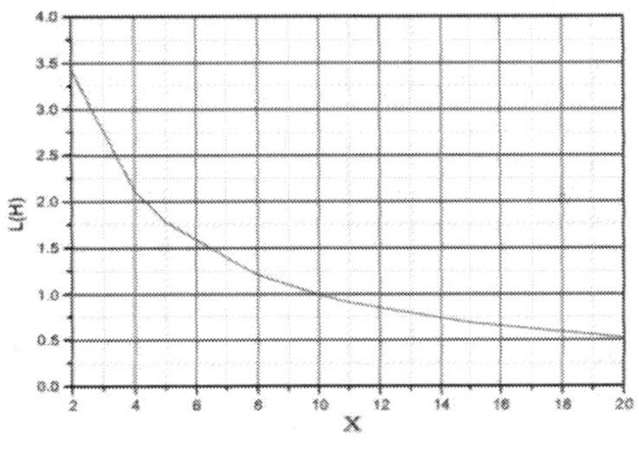

Fig. 5. The effect of sectioning on coil inductance.

3.2 Raising induction capacity via core design

Generally speaking, perm-alloy core has initial permeability μ_c up to 10000 plus, when the core is infinitely long or shaped round-loop, relative permeability μ_r is very close to μ_c. But actually core with finite length is applied rather that infinite one, when it is magnetized by outside field, polarization charge, generated in the terminal, causes the decrease of inner-core magnetic field, reduction of relative permeability μ_r, which depends on initial permeability of core and its geometric shape[6]:

$$\mu_r = \frac{\mu_c}{1+N_d(\mu_c-1)} \quad (3)$$

In the formula N_d is demagnetizing factor, related with the shape of core, μ_c being initial permeability.

$$N_d = (\frac{d}{l})^2 \bullet [\ln(2\frac{l}{d})-1] \quad (4)$$

In the formula d is diameter of core, l being its length. Generally, μ_c of core with high permeability is far more than 1, and formula (3) can be simplified as:

$$\mu_r = \frac{1}{N_d} \quad (5)$$

Through the simulation via formula (5), Figure 6 shows the relation between draw ratio and effective permeability under different initial permeability. When μ_c is up to 100000 from 1000, there is a rapid rise of μ_r, whereas μ_r is no longer raised when μ_c is up to 50000, only raised rapidly as enhanced by draw ratio [6].Therefore initial permeability of selected core shall be larger than 10000, better up to 50000, and perm-alloy initial permeability is 30000-50000, which can better fulfil the requirements.

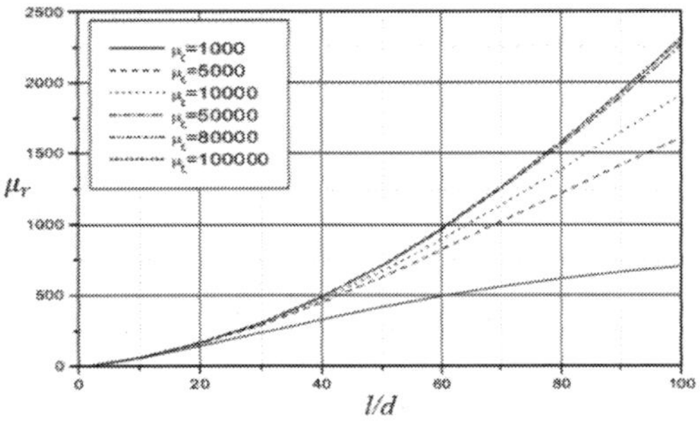

Fig. 6. The relation between draw ratio and effective permeability under different initial permeability.

Since coil length is not considered in formula (5), the calculated permeability is larger than actual value. During the design empirical formula is used to calculate effective permeability of iron core with finite length μ_e [5-6].

$$\mu_e = \frac{\mu_c\bullet[1-0.255\frac{l_c}{l}]}{1+0.975(\frac{S}{l^2})\bullet[ln(\sqrt{\frac{\pi l^2}{S}})-1]\bullet(\mu_c-1)} \quad (6)$$

In the formula l_c refers to coil length, $S=d^2$. Effective permeability via empirical formula is 30% smaller than via that formula (5), which is closer to the actual value of core material.

Table 1. μ_r and μ_e under Different l/d

l	0.25m	0.5m	1m	2m
l/d	12.5	25	50	100
μ_r	70.4	214.6	693	2326
μ_e	50	167	561	1859

μ_r and μ_e are calculated in case of l/d being 12.5-100 and d=20mm, which shows as table 1, effective permeability is risen to 3.25 times when core length is doubled.

In case of length of core being constant, relative permeability can be enhanced via reducing d as per formula (3), however what matters to the performance is the product between relative permeability and sectional area via formula (1), which needs to be raised. Figure 7 shows curve between diameter of core and $\mu_e S$ in case of length of core being 0.75m, from which parameters of magnetic sensor are raised via increasing core sectional area, with performance being decreased rapidly under d less than 2.5mm, the effect of sectional area on $\mu_e S$ being little under d more than 2.5mm. When d is started to duplicate from 2.5mm, there is only 20% rise of electrical performance. If compactness of the sensor is especially considered, core diameter can be more than 2.5mm, however when structure strength is taken into account, generally 10mm-20mm will be better. If d is to further rise, it will pose negative effect on the noise of sensor and its application since coil winding and core weight are too much [6]. Magnetic flow collector with large-area, added in both ends of core, will converge magnetic induction lines, which can easily improve effective permeability.

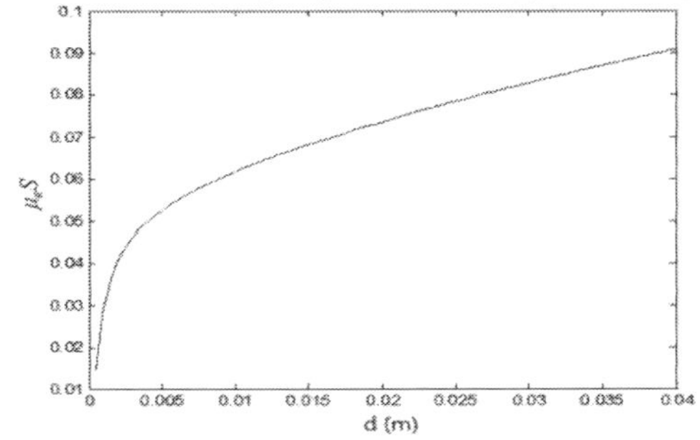

Fig. 7. Curve between diameter of core and $\mu_e S$ in case of length of core being 0.75m.

Based on the analysis above and project application, parameters are specified like core length being 1200mm, diameter 20mm, the total number of coil winding 40000, DC resistance being about 500Ω, and coil inductance 300H around. Concerned with core material, aside from initial permeability, diameter and length, there is another index being core loss, which can be reduced via wafer piling up. However, core loss resistance at 10kHz-100kHz is generally 10 folds more than that at low frequency [5], nanocrystalline with far lower loss is adopted for high-frequency application.

3.3 Low-noise pre-amplifier design

Pre-amplifier design is crucial, since it boasts the function of amplifying output voltage signal, and adjusting feedback signal strength. In light of low noise and wide band features at low frequency, it is composed of low-frequency chopping amplification channel and high-frequency amplification channel, which is shown as Figure 8. The latter consists of amplifier A1 and high-pass filter F1, directly amplifying inducted voltage via operational amplifier, whereas the former consists of modulator M1, demodulator M2, amplifier A2, low-pass filter F2, and oscillation signal source G1, amplifying the signal before modulation at fixed frequency, followed by synchronized demodulation and output. 1/f noise effect can be avoided during weak signal amplification with chopping amplification being adopted at low frequency channel, and negative feedback chopper amplifier with low-offset is applied to reduce DC offset [3]. Figure 9 shows noise curve schematic during startup and shutoff of chopping amplifier, which finds that LF noise, is apparently reduced during startup of chopping amplifier [7].

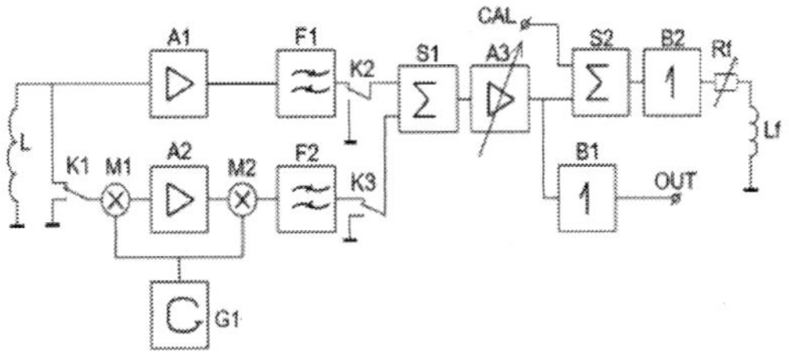

Fig. 8. Block diagram of preamplifier.

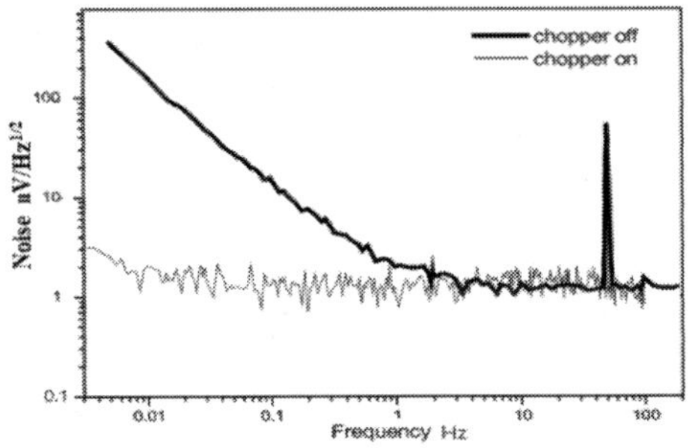

Fig. 9. Noise curve schematic during startup and shutoff of chopping amplifier.

Signals from coil *L* are paralleled into input ends of both amplifiers with same amplifying coefficient, with filter F1 and F2 boasting same cut-off frequency and series, which makes the matching and addition between S1 signal and amplifier. Total gain of amplifier is 30000 times in an effort to realize switching factor of 1.6f(V/nT). Chopping frequency is better to be 2kHz, since lower frequency breeds overall high noise of chopping amplification, and higher one results in high upwarp below 0.001Hz, formulizing new 1/f noise at super-low-frequency [7][8].

Figure 10 shows schematic block diagram of A1 and A2[2][9], in which differential amplifier with IFN860 JFET is adopted to realize low-voltage noise, low-current noise, and high-input impedance of pre-amplifier. Its input voltage is 1.8nV/Hz$^{1/2}$, and current being 10fA/Hz$^{1/2}$. Boasting the features of high-impedance and low distributive capacitance, the amplifier is suitable for wideband magnetic sensor. Noise voltage can be lower to 1nV/Hz$^{1/2}$ for amplifier with IF3602 [10], but its input distributive capacitance is high up to 300pF, which is 20 times of IFN860 fit for low frequency application, whereas not suitable for high frequency application. If power supply of ±12V is adopted for the amplifier, power loss is less than 800mW; if power supply of ±6V is adopted for the amplifier, power loss is less than 400mW.

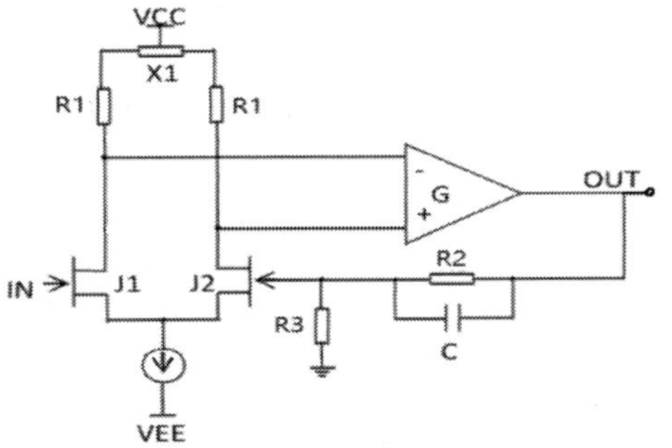

Fig. 10. Schematic block of JFET pair differential amplifier

4 Noise analysis and testing

Total noise voltage of magnetic sensor is mainly composed of heat noise of *R*, amplifier voltage noise E_n, noise of reaction between I_n and $2\pi fL$, and core loss noise R_s, of which the model is shown as Figure 11 [4][11][12].

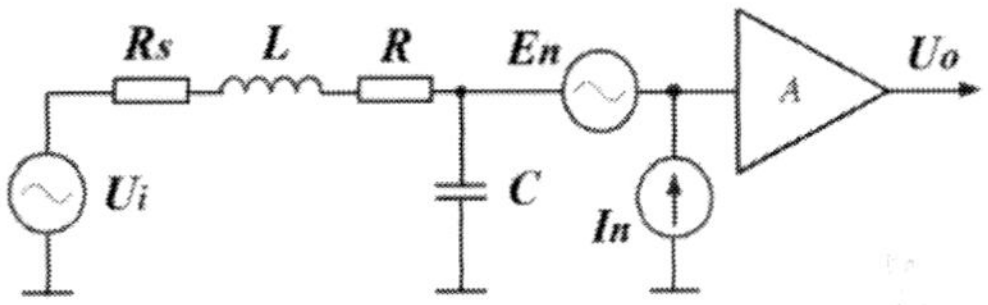

Fig.11. Noise model of magnetic sensor.

Total noise voltage U_n is:

$$U_n=(E_n^2+4kT\ (R+R_s)+(I_n*Z\ )^2)^{1/2} \qquad (7)$$

In the formula k= 1.38×10^{-23} is Boltzman constant, T=300 being absolute temperature, $Z=2\pi fL$ being coil impedance, and R_s can be neglected at low frequency.

Table 2. Calculated value of voltage and magnetic field noise

Frequency	0.01Hz	1Hz	1kHz	10kHz
I_nZ(nV/Hz$^{1/2}$)	0	0.019	18.8	188
U_n(nV/Hz$^{1/2}$)	3.4	3.4	19.1	188
B_n(nT/Hz$^{1/2}$)	6.4*10^{-3}	6.4*10^{-5}	3.6*10^{-7}	3.5*10^{-7}

Put the values into formula (7) such as heat noise of R=500Ω being 2.9nV/Hz$^{1/2}$, E_n being 1.8nV/Hz$^{1/2}$, I_n being 10fA/Hz$^{1/2}$, and L being 300H, and noise voltage at different frequencies are shown as Table 2, which shows that noise current is mainly effective above 1kHz, and noise field B_n is calculated as per general formula[11]:

$$B_n= U_n/\ (2\pi fNS\mu_e) \qquad (8)$$

When amplifier gain A=30000 times and switching factor K=1.6fV/nT are certain, B_n can be easier calculated according to $K=U_n*A/B_n$:

$$B_n= U_n*A/K \qquad (9)$$

Following sensor design, frequency response characteristic is shown as Figure 12 through actual test, with corner frequency being 0.06Hz, low-frequency switching factor being K=1.6fV/nT, and high frequency being 0.12V/nT. The comparison between noise test typical value and natural noise is shown as Figure 13, with sensor noise being 10^{-4}nT/Hz$^{1/2}$ at 1Hz, 10^{-6}nT/Hz$^{1/2}$ at 100Hz-1kHz, which is 20dB-40dB lower than natural noise, and is increased a little at 10^{-4}Hz-10^{-3}Hz owing to undesirable noise of chopping amplifier. Flat response curve, wide frequency band, and low noise features are realized for the sensor at 0.0001Hz-10kHz.

Compared with calculated value, typical value of noise test is compatible under 2kHz, whereas larger above 2kHz since lack of core high frequency loss parameter and undesirable testing environment.

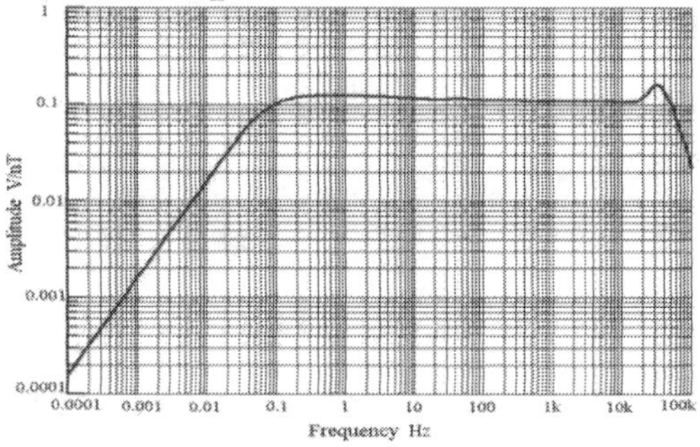

Fig.12. Frequency response feature of magnetic sensor.

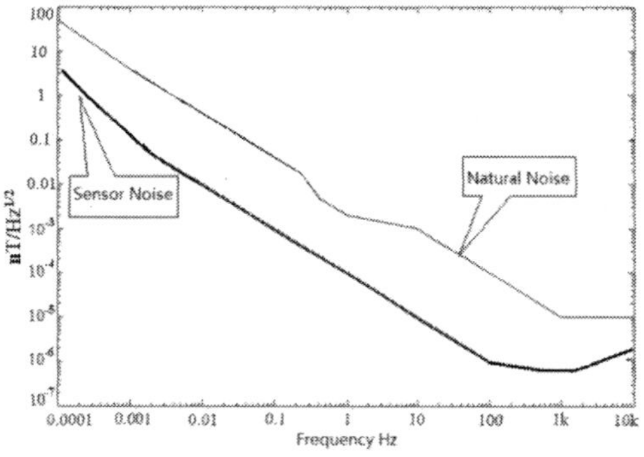

Fig.13. Comparison between typical sensor noise and natural background noise.

5 Conclusion

Coil induction capacity is enhanced through optimization of core high-permeability material selection and coil design. In order to simultaneously reach ultra-low noise and wide frequency band, this paper introduces low-noise amplifier in combination of JFET differential pair low-frequency chopping amplification channel and high-frequency amplification channel, along with magnetic feedback technology. Test result shows noise at 100Hz-1kHz can as low as to 10^{-6}nT/Hz$^{1/2}$, which reach SQUID (superconducting quantum interference device) level. Since wide frequency band 0.0001Hz-10kHz and ultra-low noise features are realized, single magnetic sensor covers operation frequencies of both MT and AMT, which can better meet the requirements of project site.

Future work will be aimed at optimizing less than 0.001Hz and higher band noise. The follow-up plan is to implement the research on the mechanism and treatment method of the extremely low noise of the chopper amplifier, and to carry out the research on the high-frequency characteristics of the high-permeability magnetic core material and the high-frequency current noise of the converter, improve the sensor design, and improve the extreme sensitivity of the sensor at extremely low and high frequency.

References

1. Y. Zhao, X. Chen, G. Wang, etc. Signal Conditioning Circuit Study of Frequency Domain Inductive Magnetic Sensor, Geophysical and Chemical Computation Technology,**3**,199-205(2016)
2. W. Huang, The Research on Compact Winding Inductive Sensor with Hollow-core, Master Degree Paper of Jilin University (2018)
3. J. Li, The Research on DC Offset Suppression of Chopper Amplifying Circuit, Master Degree Paper of Jilin University (2019)
4. X. Liu, W. Zhu, B. Yan, etc., Inductive Sensor for Lightning field measurement, Apparatus Technologies and Sensor, **1**, 5-8 (2020)
5. T. Qing,Transmission Characteristic Study of Inductive Magnetic Negative Feedback Sensor, Master Degree Paper of Jilin University (2012)
6. W. Li, Y. Liu, Y. Fan, Micro-sized Wide-frequency band Magnetic Sensor Design, Ship Electronic Engineering, **3**,129-133(2012)
7. Q. Zhao, Design and Realization of Inductive Magnetic Sensor at 0.003-10kHz, Master Degree Paper of Jilin University (2013)
8. H. Shi, Y. Wang, J. Lin, Optimal Design of Low-noise Induction Magnetometer in 1mHz-10kHz Utilizing Paralleled Dual-JFET Differential Pre-amplifier, IEEE Sensors Journal,**2537928**,1-7(2016)
9. Japheth E. Pakaree, Viranjay M. Srivastava, Realization with fabrication of double-gate MOSFET based, differential amplifier, Microelectronics Journal **91** (2019)
10. K. Liu , Ultralow Noise Preamplifier and Optimization Method for Induction Magnetometers, IEEE SENSORS JOURNAL, **VOL.15**,3293-3300(2015)
11. A. Grosz, E. Paperno,A. O. of L. S. C. M., IEEE SENSORS JOURNAL, **VOL. 12**,2719-2723(2012)
12. M. Abdul Mubin, A. Marzuki, A Low-Noise Amplifier Utilizing Current-Reuse Technique and Active Shunt Feedback for MedRadio Band Applications, International Journal of Electrical and Electronic Engineering & Telecommunications (2020)

MATEC Web of Conferences **327**, 01003 (2020) https://doi.org/10.1051/matecconf/202032701003
ICMIE 2020

Glucose Analyzer Based on Self-made Biosensor for High-performance Glucose Detection

Qiang Tan[1], *Cuimei* Bo[1,*], *Jun* Li[1,*], *Yiqing* Wang[1], *Xiaorong* Wang[1], *Shubo* Jiang[1]

[1] College of Electrical Engineering and Control Science, Nanjing Tech University, Nanjing, China

Abstract. The concentration analyzer with high precision and wide range is the core device for monitoring the fermentation process. In this work, we designed and proposed a lowcost three-electrode glucose analyzer based on a self-made screen-printed enzyme biosensor chip, which has a Prussian blue (PB) nanocubic structure and leads to high sensitivity of 117.31 $\mu AmM^{-1}cm^{-2}$. The hardware design of the glucose analyzer can be divided into five critical parts, including digital, signal treatment system, power supply, motor-driven and the host computer. The signal treatment system is used to collect, convert and amplify the weak current signal generated by the biosensor. The digital circuit of the central processing unit of the analyzer is designed using the STM32F407ZET6. Besides, an external analog-to-digital converter is used to achieve high precision A/D conversion. The stability of the potentiostat is ensured by designing the precision power supply, hardware filtering, and algorithm filtering. The experimental results show that the glucose analyzer has a wide linear detection range from 1g/L to 120g/L and the coefficient of variation at 1g/L is 0.038, which exhibits excellent performance in stability and detection accuracy. The analyzer can be applied in the future for in-situ measurement of glucose concentration for its wide-range and high-precision detection capabilities.

1 Introduction

In order to obtain high efficiency with low cost, the continuous operation dominates the industrial operation in large-scale manufacturing, which strongly relies on the real-time control of the microbial reaction to realize an optimum transfer rate. Therefore, online monitor technology is the key to guide the operation by receiving dynamic fluctuation of the essential parameters[1]. Currently, there have been already some parameters showing the environmental state in the fermentation reactor can be online detected, such as temperature, pH value, pressure and concentration of dissolved oxygen[2]. However, the real-time detection of constituent concentration, such as glucose, which is directly served as a common substrate to associate with the reaction kinetics in fermentation, has been paid rare attentions to the development of testing instruments.

In recent decades, the application of enzyme biosensors in biosensing has been widely investigated and achieved significant progress[3-5]. Due to the inherent substrate specificity of enzymes and their ability to operate under mild conditions, enzymes have become a biocatalytic pathway, which has caused great research interest [6]. The enzyme biosensor is made by combining a biological enzyme with various electrochemical electrodes or directly fixing the biological enzyme on a substrate electrode[7]. The glucose enzyme biosensor [8-11] have become essential analytical tools in the fields of clinical analysis, environmental monitoring and food safety due to their fast response, high accuracy and low cost [12].

In enzyme biosensors, the properties of electrolytes, membranes, electrodes, and microorganisms change over time[13]. Besides, to achieve high-precision applications and be able to process tiny signals, the noise should be controlled at an extremely low level. In this work, a lowcost three-electrode glucose analyzer based on a self-made screen-printed enzyme biosensor chip is developed, which includes the design of signal treatment system, digital circuit, precision power supply circuit, motor drive circuit and the host computer. The reliability and stability of the system can be ensured by the precise power supply and stable detection environment. The response current of the enzyme electrode is decreased after long-term use, the system automatically calibrates and establishes the concentration-response equation to eliminate the error. The experimental results show that glucose analyzer has a wide linear detection range.

2 self-made biosensor chip

In previous work by our team [14-16], PB enzyme biosensor chips with nanocubic crystals were manufactured in batches by screen printing technology for the general detection of various physiological substances. As shown in Fig.1, two solutions were prepared as follows: solution A, 0.01 M $K_4[Fe(CN)_6]$ +

* Corresponding author: lj_bcm@163.com

 https://doi.org/10.1051/mateconf/202032701003

0.1 M KCl + 0.1 M HCl, and solution B, 0.01 M FeCl3 + 0.1 M KCl + 0.1 M HCl. The syringe was fixed to the automatic syringe pump at a rate of 1000 L/min after solution A and solution B were poured into the syringe. At the same time, a beaker at the heating magnetic stirrer was used for reaction vessel, in which the beaker contained distilled water before the start of the reaction. A constant temperature water-bathing was used to obtain the most suitable nanostructure of PB. The PB slurry was obtained by centrifuging PB solution at the beaker, and the mixed ink was formed by mixing the PB slurry with the carbon ink. Then, the PB slurry was screen-printed as a working electrode(WE), a counter electrode(CE) and a connector were generated using carbon ink, a reference electrode (RE) was printed using silver chloride ink. Commercial PVC boards were invoked as the substrate for screen printing. A glutaraldehyde cross-linking method was utilized to fix glucose oxidase (GOD) on the working electrode, and a three-electrode enzyme biosensor chip was finally fabricated.

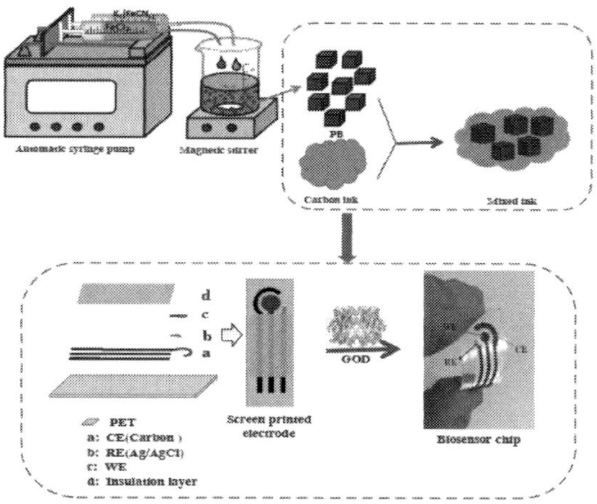

Fig.1.Preparation of biosensor chip

The excellent performance of the self-made biosensor chip has been proved by a series of tests. Fig.2(a) showed the CV diagrams of the biosensor, and the scan rate was 0.05 V s^{-1}, hundredth redox peak current is still able to achieve 94.5% of the first, indicating that the chip has a very excellent repeatability. Fig.2(b) showed the interference of ascorbic acid (AA) and uric acid (UA) on the responses of the glucose biosensor in 0.05 M PBS containing 0.1 M KCl. Fig.2(c) showed the sensitivity of 10 chips randomly selected for glucose. Fig.2(d) showed the sensitivity changes over time, and the chip is tested for performance every 5 days and stored at 4°C in air. The self-made enzyme biosensor chip has excellent selectivity, reproducibility and stability at a low operating potential of -0.05V.

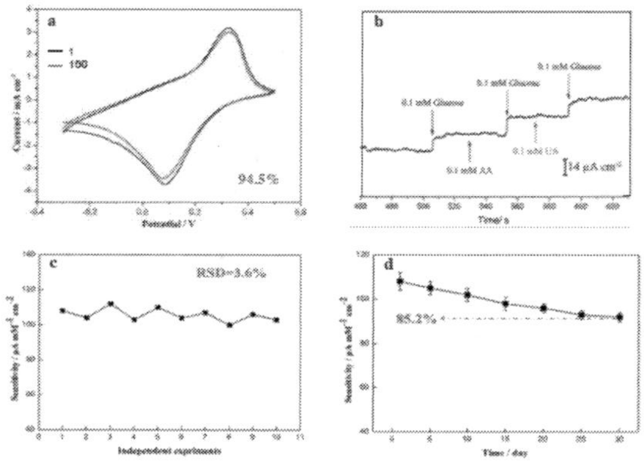

Fig. 2. Performance tests of self-made biosensor chip.

3 Design of analyzer

3.1 hardware design

Hardware can be divided into five parts including digital design, power supply, motor-driven, the host computer and signal treatment system,. The first is the power supply, the stability of the power supply is under a massive impact on the operation of the entire analyzer. Second, the digital design is mainly based on the STM32F407ZET6 microcontroller unit (MCU), which has the functions of data storage and processing, control of the stirring motor and communication. ADS8332 chip is utilized to achieve 16-bit high-precision analog-to-digital conversion. Followed by the design of the motor drive circuit. The glucose measurement needs to be undertaken in a test cell equipped with PBS. A magnetic stirring motor is required to mix PBS and glucose solution; discharge the waste solution after the detection is completed. Fourth is the host computer for human-computer interaction. RS-485 circuit is designed for communication between the host computer and the MCU. The host computer can display all the information in the detection process and control the glucose analyzer. The most critical signal treatment system will be described in detail in the next section. As shown in Fig.4.All hardware is printed into a circuit board as a more stable and long-lasting solution.

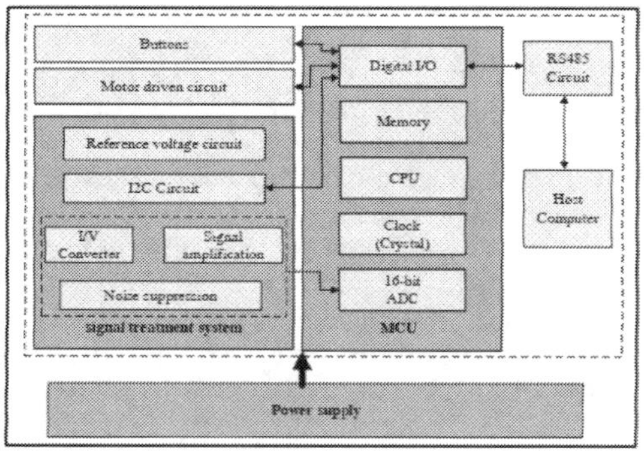

Fig.3. Hardware block diagram.

3.2 Signal treatment system

In a three-electrode system, the RE provides a reference voltage and the working WE promote the electrochemical reaction of the measured solution under the effect of the external excitation signal and then, measures the current formed on the electrode. Meanwhile, the CE forms a circuit with the WE. The measured solution is excited by a voltage signal to produce a redox reaction and generate a weak current. The value of the formed current depends on the concentration of the measured substance. Hence, the concentration of the measured substance can be calculated based on this relation. When performing electrochemical detection, a constant potential should be added between WE and RE to maintain the electrochemical stability of the biosensor.

We used an electrochemical analog front-end chip to build a signal treatment system for this glucose analyzer. The schematic of the simplified circuit is shown in Fig.3. The internal control amplifier A1 of the LMP91000 constitutes a potentiostatic circuit, and the error signal is amplified and fed back to the CE of the biosensor. The impedance change between the WE and the RE will cause the changes in the voltage of CE, which ensures the stability of the voltage between the RE and the WE. The output voltage of the transimpedance amplifier(TIA) connected to the WE is proportional to the working current of the biosensor. Meanwhile, the WE is connected to the virtual ground (internal ground) through the TIA. The MCU communicates with the LMP91000 through the I2C interface, which can change the gain of the internal transimpedance RTIA and can also select the reference voltage and bias voltage of the constant potential amplifier. The I2C interface, which can change the gain of the internal transimpedance R_{TIA} and can also select the reference voltage and bias voltage of the constant potential amplifier.

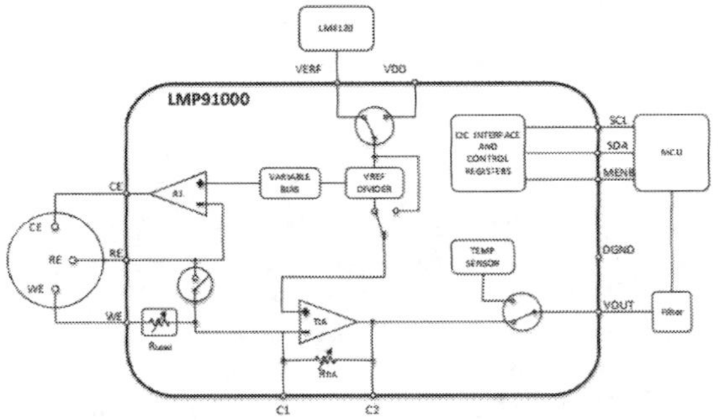

Fig.4. Preparation of biosensor chip

3.3 Algorithm filtering

High-precision 16-bit ADC was used for analog-to-digital conversion and the ADC sampling period was set to 10ms. A high-performance first-order inertial filtering was adopted to filter out the interference of random signals. The formula is as follows:

$$Y(n) = \frac{\alpha X(n) + (K-\alpha)Y(n-1)}{K} \qquad (1)$$

where K is the magnification of the coefficient α, α is an integer from 0 to K, X(n)is the actual AD value collected this time, Y(n-1)is the AD value obtained after the last filtering, and Y(n) is the AD value obtained through this filtering.

In terms of the first-order inertial filtering, we used the AD value collected this time to weight the AD value received last time to obtain a valid filter value, which leads to the previous output signal can effectively be fed back to the next signal. The algorithm filtered out the interference with the tiny signals in the system and obtained accurate AD values related to the glucose concentration, despite a delay of about 1s in the acquisition of concentration information. As presented in Fig.5. Different values of a are used to compare the filtering effect, the sampling period of the ADC was 10ms. The AD value collected during a test was converted into a corresponding voltage value (mV). First-order inertial filtering was used for simulation in MATLAB. K was 256.

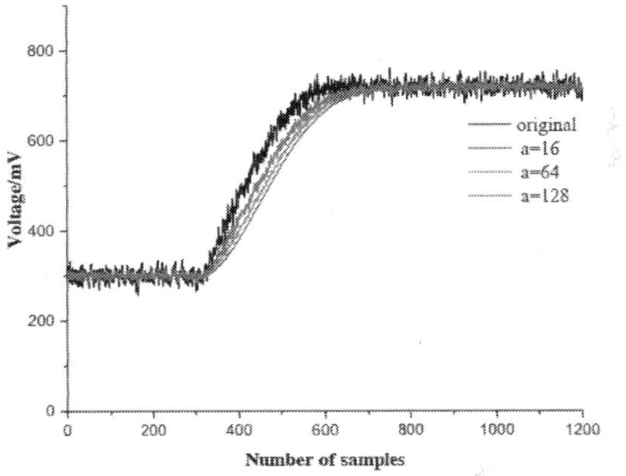

Fig.5. Test of first-order inertial filtering. a is α of formula 1.

3.4 Calibration

The sensitivity of the biosensor chip will decrease with time, and the response current will decrease. This requires the system to calibrate and establish a concentration-response equation to eliminate the error. The glucose concentration is proportional to the response current (or voltage) in the linear detection range of the electrochemical enzyme biosensor. Thus, a known concentration of a standard glucose solution can be used to calibrate this electrochemical enzyme biosensor, which can be explained by the following concentration-response equation:

$$y = \frac{ADV_0 - ADV_1}{C_0} x + ADV_0 \qquad (2)$$

where ADV_0 is the AD value of the solution injected into the PBS which contains no glucose, ADV_1 is the AD value of the standard glucose solution of known concentration injected into PBS, C_0 represents the concentration value of the standard glucose solution, y is the AD value of the solution injected into the PSB of which the glucose concentration is unknown, and x is the glucose solution concentration to be measured.

4 Experiments and results

At room temperature, the biosensor chip was placed in a test cell with 3 mL of PBS solution in it. As presented in Fig.6. 5 uL of PBS buffer was injected into the test cell to obtain ADV0 after turning o n the machine and activating the biosensor chip. Secondly, drained the waste solution from the test cell, cleaned the test cell, inject 3mL of PBS solution again, and finally injected 5uL of glucose solution (60g/L) to obtain the value of ADV1. The concentration characteristic equation was established according to the calibration formula, and thus the calibration of the biosensor chip was completed. To ensure the calibration was correct, we repeat the second step of injecting 5uL of glucose solution (60g/L). If the testing result is 2% below or above the 60g/L concentration, the calibration is successful. Otherwise, recalibrate.

When the calibration finished, a linearity test was conducted by using a glucose solution of known concentration. The measured increment was: 1) 0.2g/L under the concentration from 1 to 2g/; 2) 2g/L under the concentration from 2 to 10g/L; 3) 10g/L from 10g/L to 110g/L; and 20g/L form 110 to 120g/L. Each increment was tested five times and averaged. Electrolytes, membranes, electrodes and microorganisms change their properties overtime during the test. In this work, we recalibrated every hour to eliminate this error. The final result is shown in Fig.7.

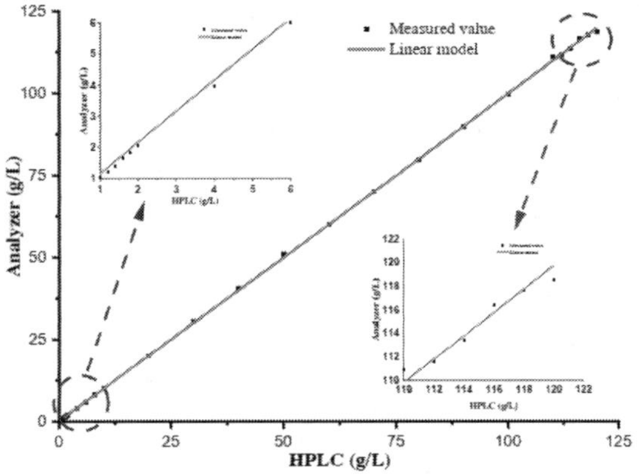

Fig.7. Linearity test of glucose analyzer

The linear regression equation for calculating the concentration (x) measured by HPLC and the concentration (y) measured by the instrument, y = 0.9967x + 0.1708, R2 = 0.999. The experimental results showed that the analyzer has good linearity between the glucose concentration of 1-120g/L. The analyzer fully utilizes the linear measurement capability of the self-made enzyme biosensor. The motor had been rotating at an appropriate speed in the process of measurement to prevent the start and stop of the stirring motor from affecting the detection environment.

1g/L and 60g/L were selected for stability testing which was shown in Fig.8. Each concentration was tested three times a day for 21 consecutive days with other conditions unchanged. In the continuous stability test, the analyzer and the self-made biosensor showed outstanding stability. 63 tests were carried out on 1g/L glucose and the average measured concentration was 0.988g /L, the coefficient of variation was 0.038. The analyzer has a high measurement accuracy even at low levels. Meanwhile, 60g/L was used for testing and the mean measured concentration was 59.57g/L, the coefficient of variation was 0.01.

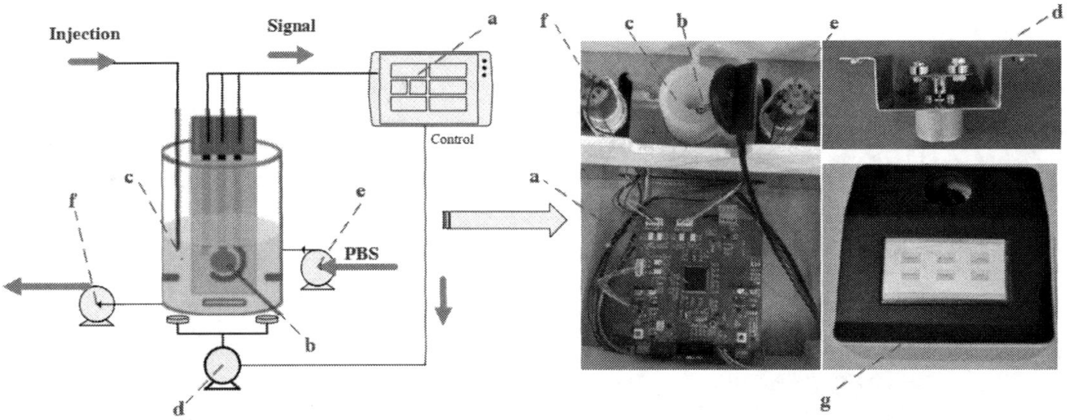

Fig.6. Analyzer structure. (a) Hardware, (b) self-made glucose biosensor chip, (c) test cell, (d) Magnetic stirring motor, (e) Inlet pump, (f) Outlet pump, (g) Overall image of the instrument. Each time Before testing, wash the test cell with PBS, pump 3mL PBS through the peristaltic pump, and then inject glucose solution for concentration detection. After the test, the Waste solution is discharged through the pump.

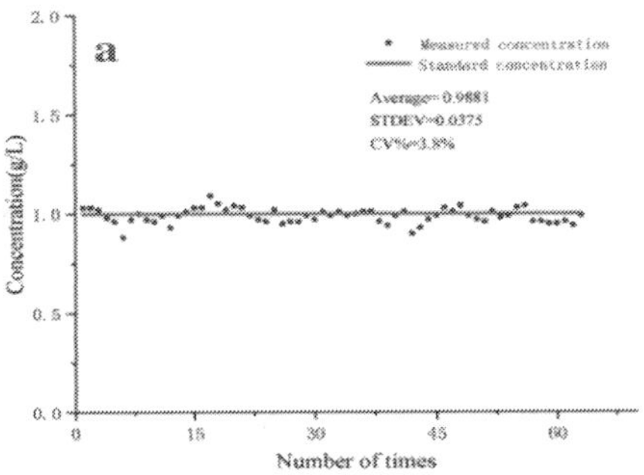

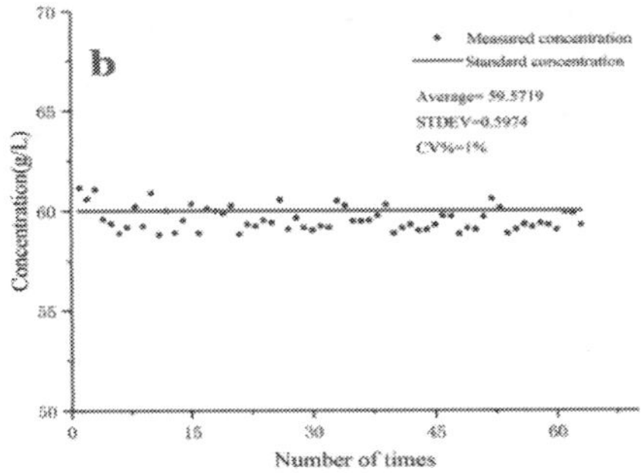

Fig.8. Stability test of glucose analyzer. (a)test with 1g/L of glucose (b) test with 60g/L of glucose

5 Conclusions

In this work, we designed a glucose analyzer based on self-made biosensor chips. The analyzer has a wide linear detection capability of 1-120g/L in the detection environment of PBS; the CV at 1g/L is 0.038. The glucose analyzer has excellent stability and measurement accuracy. This glucose analyzer is quite promising for in-situ on-line detection of glucose concentration under various conditions, especially in the fermentation process.

Acknowledgements

This work was financially supported by National Natural Science Foundation of China (No. 2172780038), open subject of State Key Laboratory of Materials Chemistry Engineering ().

References

1. Lisa Mears, Stuart M. Stocks, Gürkan Sin, Krist V. Gernaey, *A review of control strategies for manipulating the feed rate in fed-batch fermentation processes*, J. Biotechnol, **245**, 34 (2017)
2. Xuzhi Zhang, Xiaoyu Jiang, Zhihui Hao, Keming Qu, *Advances in online methods for monitoring microbial growth*, Biosens. Bioelectron, **126**, 433(2019)
3. Elif Burcu Bahadır, Mustafa Kemal Sezgintürk, *Applications of commercial biosensors in clinical, food, environmental, and biothreat/biowarfare analyses*, Anal. Biochem, **478**, 107 (2015)
4. L.C. Clark, C. Lyons, *Electrode systems for continuous monitoring in cardiovascular surgery*, Ann. N. Y. Acad. Sci, **102**, 29 (1962)
5. Conan Mercer, Richard Bennett, Peter Ó. Conghaile, James F. Rusling, Dónal Leech, *Glucose biosensor based on open-source wireless microfluidic potentiostat*, Sens. Actuators B， **290**, 616 (2019)
6. J. Wang, *Electrochemical detection for microscale analytical systems: a review*, Talanta, **56**, 223 (2002)
7. F. Arduini, C. Zanardi, S. Cinti, F. Terzi, D. Moscone, G. Palleschi, R. Seebe, *Effective electrochemical sensor based on screen-printed electrodes modified with a carbon black-Au nanoparticles composite,* Sens. Actuators B, **212** , 536 (2015)
8. S. Calabrese Barton, J. Gallaway, P. Atanassov, *Enzymatic biofuel cells for implantable and microscale devices*, Chem. Rev **104**, 4867 (2004)
9. D. Leech, P. Kavanagh, W. Schuhmann, *Enzymatic fuel cells: recent progress*, Electrochim. Acta, **84**, 223 (2012)
10. J. Wang, *Carbon-nanotube based electrochemical biosensors: a review*, Electroanalysis, **17**, 7(2005)
11. Ievgen Mazurenko, Vivek Pratap Hitaishi, Elisabeth Lojou, *Recent advances in surface chemistry of electrodes to promote direct enzymatic bioelectrocatalysis*, Current Opinion in Electrochemistry **19**, 113 (2020)
12. Xiuling MAO, Jian WU, Yibin YING, *Application of Electrochemical Biosensors in Fermentation*, Chin. J. Anal. Chem. **36**, 1749(2008)
13. Turner, Robin FB, D. J. Harrison, and HENRY P. Baltes, *A CMOS potentiostat for amperometric chemical sensors*, IEEE J. Solid-State Circuits, **22.3**, 473 (1987).
14. Lifang Liu, Lei Shi, Zhenyu Chu, Jingmeng Peng, Wanqin Jin, *Prussian blue nanocubes modified graphite electrodes for the electrochemical detection of various analytes with high performance*, Sens. Actuators B, **202**, 820 (2014)
15. Danfeng Jiang, Zhenyu Chu, Jingmeng Peng, Wanqin Jin, *Screen-printed biosensor chips with Prussian blue nanocubes for the detection of physiological analytes*, Sens. Actuators B, **228**, 679 (2016)
16. Linlin Li, Jingmeng Peng, Zhenyu Chu, Danfeng Jiang, Wanqin Jin, *Single layer of graphene/Prussian blue nano-grid as the low-potential biosensors with high electrocatalysis*, Electrochimica Acta, **217**, 210 (2016)

An ISFET sensor based on In_2O_3 nanoribbon for pH detection of micro solutions

Chuanjian Wu[1], *Yiqing* Wang[1, *], and *Min* Yang[1]

[1] College of Electrical Engineering and Control Science, Nanjing Tech University, Nanjing, China

Abstract. In this paper, an ISFET sensor based on In2O3 nanoribbon is developed for pH detection of micro solutions. A double channel ISFET is designed which includes the In2O3 nanoribbon top channel and silicon substrate bottom channel so as to realize the pH detection of the micro solutions. Through experiments, the sensitivity of the sensor is 70 mV/pH in the range of pH 6 to pH 10, which exceeds the Nernst limit of 59.2 mV/pH (300 K) at room temperature. Taking the detection of cTnI as an example, the pH sensor can be used to detect the concentration of biomarkers.

1.Introduction

The pH detection has important applications in industry, biomedicine, environmental protection and so on [1]. The test strip and the glass electrode are widely used pH sensors. However, the subjective reading of the pH test paper is not accurate. The glass electrode is fragile, and is difficult to further miniaturization. Moreover, the glass electrode is particularly insensitive to strong acid or alkali [2]. Therefore, it is urgent to develop a new material to detect the pH value of the micro solution which may realize the miniaturization. In 1970, P. bergveld first proposed the ISFET structure for the detection of pH (H^+ ion) value [3]. Compared with the traditional pH sensors, the ISFET has more advantages in durability, rapidity, accuracy, sensitivity and miniaturization.

In recent years, the addition of biocompatible nanomaterials to the gate region of ISFET biosensors has attracted extensive attention of more and more researchers [4]. ISFET sensors based on silicon nanowires, zinc oxide nanowires, carbon nanotubes, graphene and other nanomaterials are spread to apply for the detection of pH value, the protein, the DNA and other biological substances [5-8].

In this paper, we propose a nanostructured ISFET device to detect the pH value of the micro solutions. Rest of the paper is organized as follows. The operation mode of the sensor and the measurement results are explained in detail in section 2. The structure and pH performance of ISFET are analyzed in section 3. Some conclusions are given in the conclusion part.

2.Materials and method

2.1 Detection principle

The sensitive mechanism of ISFET is based on surface base theory and double layer theory [9-10]. According to the surface basis theory, there are an amount alcohol groups (-OH) on the surface of the insulator, which can be

protonated (positively charged) or deprotonated (negatively charged) depending on the concentration of hydrogen ions in the electrolyte. The phenomenon may induce the Nernst response potential which is sensitive to H^+ at the interface between the sensitive membrane and electrolyte solution, so that the current between source and drain may be regulated.

According to the theory of electric double layer, there exists ionic charges on the surface of insulating layer and the electrolyte. The charges distributed on the surface of insulating layer may inevitably attract the same amount of electric charges in the electrolyte, forming an anti-ion layer. Hence, an electrical double layer between the insulating layer and the electrolyte interface has the same amount of the electricity, and opposite polarity. Therefore, a potential difference is generated between the surface of the insulating layer and the electrolyte solution, which is called the interface potential E. The relationship between the interfacial potential and the logarithm of H^+ activity in electrolyte solution is linear and it is expressed as,

$$E = E_0 + \frac{RT}{F} \ln a_{H^+} \ , \quad (1)$$

where R, T, Tre, F, a_{H^+}, E_0, respectively represents the gas constant, the absolute temperature, the Faraday constant, the activity of hydrogen ion and a constant.

The dielectric layer is usually very thin, and the interface potential E may change the charge density on the semiconductor surface, which may further regulate the source leakage current. The relationship between source leakage current and interface potential is written as follows,

$$I_{DS} = \frac{\mu_n C_o W}{L}\left[\left(V_{GS} - E_{ref} + E - V_T\right)V_{DS} - \frac{1}{2}V_{DS}^2\right], \quad (2)$$

* Corresponding author: wangyiqing@njtech.edu.cn

https://doi.org/10.1051/matecconf/202032701004

where μ_n, C_o, V_{GS}, V_{DS}, w/L, V_T, E_{ref} respectively stand for the carrier mobility, the gate capacitance per unit area, the gate source voltage, the drain source voltage, the channel width length ratio, the threshold voltage, and the reference electrode and solution interface potential.

Combining Eq. (1) and (2) may get following relation,

$$I_{DS} = \frac{\mu_n C_o w}{L}\left[\left(V_{GS} - E_{ref} + E_0 + \frac{RT}{L}\ln a_{H^+} - V_T\right)V_{DS} - \frac{1}{2}{V_{DS}}^2\right]. \quad (3)$$

As I_{DS}、E_{ref}、V_{DS} are all constant, we have the following relation,

$$V_{GS} = A + B\ln a_{H^+} \ . \quad (4)$$

As V_{GS}、E_{ref}、V_{DS} are all constant, we may get,

$$I_{DS} = C + D\ln a_{H^+} \ . \quad (5)$$

From Eq. (4) and (5), the relationship between the gate source voltage and the source leakage current is linear with logarithm of the H^+ activity in solution. The pH value of the solution may be acquired through the measurement of the hydrogen ion activity in the solution by the gate source voltage or the source leakage current.

2.2 In_2O_3 nanoribbon ISFET

In_2O_3 has stable chemical properties, wide band gap and low resistivity. Applying the field effect transistor with nanostructured In_2O_3 to the channel material has the advantages of high signal-to-noise ratio, label free and easy to multi-channel integration. Nanomaterials can be considered as one-dimensional. As one-dimensional metal acts as a "flexible" electron channel, the atomic stress within or between molecules may lead to the change of electrochemical behavior several orders of magnitude larger than that of bulk materials [11]. Using one-dimensional FET for biochemical sensing may greatly improve the sensitivity of the device [12].

The structure of In_2O_3 nanoribbon ISFET is showed in Fig. 1. The length and width of the device are respectively 15 mm and 10 mm. Si_3N_4 is used as the pH sensitive film material which is solidified on silicon substrate. The nanoribbon is made by In_2O_3 material and Ti/Au is applied to make the source and drain. The nanostructured field-effect transistors (FET) is fabricated by plasma enhanced chemical vapor deposition (PECVD), electron beam evaporation, photolithography and magnetron sputtering.

The preparation process of In_2O_3 nanoribbon ISFET is introduced as follows. Firstly, the Si_3N_4 film is grown on Si substrate by the plasma enhanced chemical vapor deposition (PECVD). Secondly, source and drain are fabricated by one-step photolithography, and source and drain masks are formed by coating photoresist on Si_3N_4 film. Thirdly, Ti/Au metal is deposited by electron beam evaporation, and the source and drain are generated after stripping the metal. Fourthly, the gate is fabricated by secondary lithography. Then coating the gate with photoresist may form a gate mask. Finally, the In_2O_3 nano film is deposited by magnetron sputtering which may result a channel. It has a growth/width ratio of 800μm/50μm and a thickness of 50nm. The grid is formed after stripping coats.

Fig.1. The structure of ISFET

2.3 Micro Device

In order to make the electrolyte solution fully contact with the ISFET surface of In_2O_3 nanoribbon, a solution filling device with a capacity of 2 mL is designed in this paper. The overall structure of the equipment is showed in Fig. 2. A cylinder with a diameter of 3.2cm and a height of 1.5cm is embedded with a hollow conoid with a diameter of 1.5cm and a height of 2.2cm. The bottom of the conoid is a circle with a diameter of about 2 mm and an outer wall of about 1 mm, which can very well fit the surface of ISFET. The cylinder is fixed by four screws. All of tests have been performed by the device.

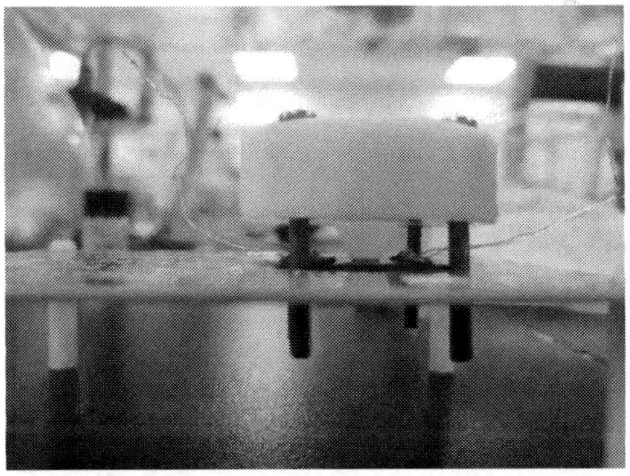

Fig.2. Micro detection cell

2.4 Measurement setup of In_2O_3 nanoribbon ISFET

Aiming at studying the electrical and pH characteristics of nanostructured ISFET, the output characteristic curve (I_{DS}-V_{DS}) and transfer characteristic curve (I_{DS}-V_{GS}) are detected by our own detection circuit. All the detected liquid gate potentials are applied through the Ag/AgCl reference electrode. The output and transfer characteristics are measured at room temperature and in 0.1 * PBS buffer solution. We use DP1308A (Beijing Puyuan Jingdian Technology Co., Ltd.) to supply a fixed potential of -1V to the liquid gate, and scan the voltage from -4V to 0V between the source and drain to measure the output characteristics. The transfer characteristics are measured under fixed -0.5V source drain voltage, and the liquid gate potential is scanned from 0V to 2V.

For the purpose of detecting the pH sensitivity, the I_{DS}-V_{GS} curves of In_2O_3 nanoribbon ISFET are measured at room temperature using commercial standard pH solutions with pH values ranging from 6 to 10. The liquid gate voltage is scanned from 0V to 1.3V, and the source drain voltage is fixed at -1V.

3. Results and discussion

The output characteristic curve and transfer characteristic curve of In_2O_3 nanoribbon ISFET are illustrated in Fig. 3 and Fig.4. Firstly, the In_2O_3 nanoribbon ISFET biosensor is prepared by aforementioned method. Then, after the chip is placed on the electrochemical stage, the gate and source drain electrodes are connected to detect the electrical signal. As 1.5ml solution is added, due to the existence of H^+ in the solution and H^+ formed by hydration of In_2O_3 film surface, a double electric layer is formed at the interface of one side of the solution. Thus, an interface potential is generated. The formed interface potential changes the charge density of the In_2O_3 surface and generates the interface potential difference, which regulates the channel current. As different grid voltages are applied, the interface potential may correspondingly vary. The high gate voltage results the small interface potential difference, and the small source leakage current.

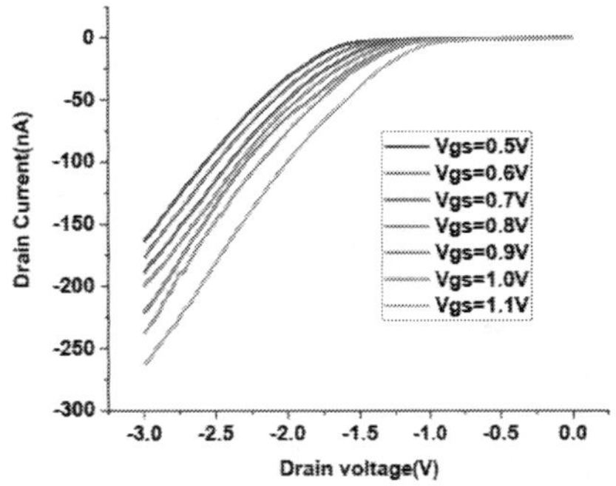

Fig.3. The output characteristic of nano structure based ISFET

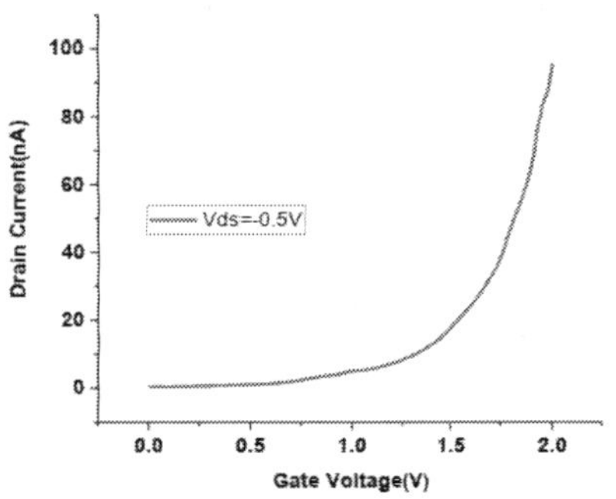

Fig.4. The transfer characteristic of nano structure based ISFET

The voltage between the source and drain is set to -1V, since the ISFET device is in deep saturation. In the saturated region, the current (I_{DS}) between source and drain is weakly dependent on the voltage between source and drain (V_{DS}), which is mainly controlled by the voltage between gate and source. Even if the V_{GS} change very tiny, the I_{DS} may significantly increase. As shown in Fig. 5, a family of I_{DS}-V_{GS} curves is displayed with pH values from 6 to 10. As adding solutions with different pH values, the current obtained is changed since the H^+ concentration is different. From the Fig. 6, it can be seen that the I_{DS} is enhanced with increasing of pH value.

At 10μ a drain source current, the sensitivity of In_2O_3 nanoribbon ISFET is 70±5mV/pH, which exceeds the Nernst limit at room temperature. The sensitivity structure is presented in Fig. 5. Compared with the traditional ISFETs, the advantage of the proposed device is obvious that our device is not only has semiconductor silicon channel, but also has In_2O_3 nanoribbon acting as conductive channel. Under dual channel operation, the pH sensitivity of the device easily exceeds the Nernst limit. From results, it can already achieve the desired effect in the detection of biomaterial concentration depending on the change of pH value.

On the basis of the successful detection of pH value by In_2O_3 nanoribbon ISFET sensor, we tried to detect cTnI at 100 pg/ml. Enzyme linked immunosorbent assay (ELISA) is used to detect the source leakage current of cTnI at 100 pg/ml through fixed liquid gate potential and source drain voltage. Results of tests are investigated in Fig. 7. Firstly, taking 0.5ml 0.1*PBS solution, the measured source leakage current is stable at -0.5294 mA. Then, after we add 1ml of 1mg/ml urea solution, the measured source leakage current is stable at -0.4126 mA. The change of cTnI concentration in solution makes the change of the surface potential of ISFET, resulting to the difference of source leakage current. The variation of source leakage current reflects the concentration of different cTnI.

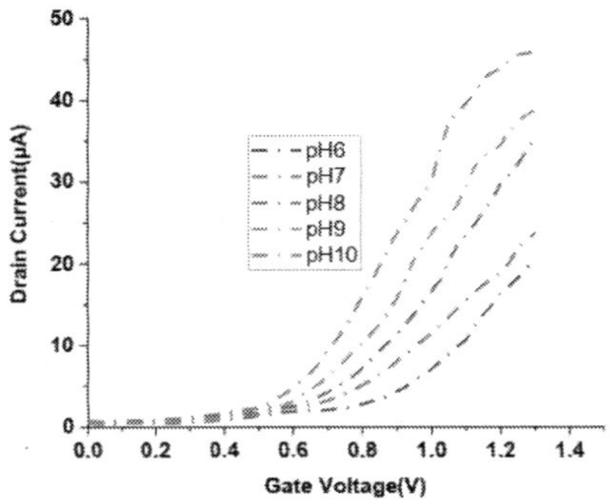

Fig.5. The family of I_{DS}-V_{GS} curves with pH values from 6 to 10

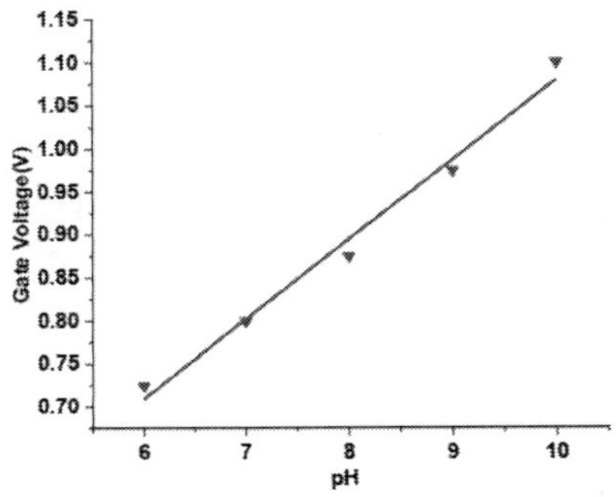

Fig.6. The linearity of pH values from 6 to 10

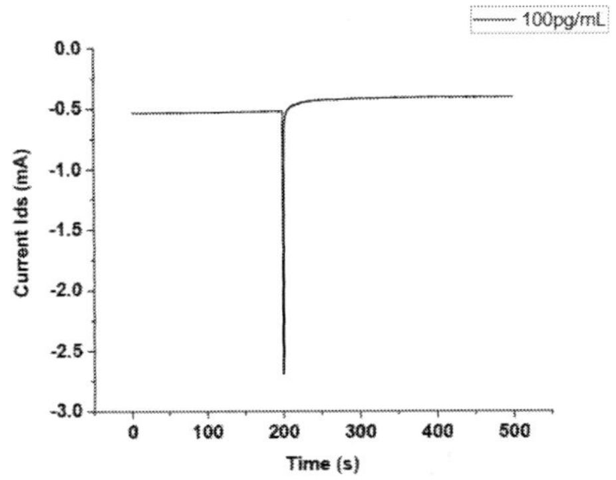

Fig.7. The results of 100pg/mL cTnI by In_2O_3 Nanoribbon ISFET

4. Conclusion

A kind of pH sensor based on In_2O_3 nanoribbon ISFET is introduced for the detection of pH value of micro solution in this paper. Through dual channel operation of the proposed ISFET, the device performs high pH sensitivity, reaching at 70 mV/pH. In_2O_3 nanomaterials have good biocompatibility so that it is widely used to detect biomolecules based on pH detection, such as cTnI, NMP22 protein of bladder cancer, amyloid beta peptide of Alzheimer's disease. Therefore, the proposed device has a very good application prospect.

5. Acknowledgement

This study was funded by the Jiangsu Province Natural Science Foundation for the Youth (No. BK20180687).

Reference

1. P. Salvo, B. Melai, N. Calisi, et al. *Graphene-based devices for measuring pH*, Sens. Actuators B,256(MAR.), pp.976-991,(2018).
2. S. Karastogianni, S. Girousi, S. Sotiropoulos, *pH: principles and measurement*, Encycl. Food Health, pp.333–338(2016).
3. P. Bergveld. *Development of an ion-sensitive solid-state* device *for neurophysiological measurements*, IEEE Transactions on Bio-Medical Engineering,**17(1)**, pp.70-71, (1970).
4. A. Nehra, K. Pal Singh, *Current trends in* nanomaterial *embedded field effect transistor-based biosensor*, Biosens. Bioelectron,**74**, pp.731-743, (2015).
5. K. Zhou, Z. Zhao, P. Yu, Z. Wang, *Highly sensitive pH* sensors *based on double-gate silicon nanowire field-effect transistors with dual-mode amplification*, Sens. Actuators B,**320**, (2020).
6. K. Sun, I. Zeimpekis, C. Hu, N.M.J. Ditshego, O. Thomas, M.R.R. de Planque, H.M.H. Chong, H. Morgan, P. Ashburn, *Low-cost top-down zinc oxide nanowire sensors through a highly transferable ion beam etching for healthcare applications*, Microelectronic Engineering,**153**, pp.96-100, (2016).
7. M. Sarma, P. Sharma, J. Chandra Dutta, *Modelling of Potassium-Doped Polypyrrole/Carbon Nanotube-Based Enzyme Field Effect Transistor for Cholesterol Detection*,Materials Today: Proceedings,**4**,9, pp.10346-10350,(2017).
8. Cao Thi Thanh, Nguyen Hai Binh, Nguyen Van Tu, et al, *An interdigitated ISFET-type sensor based on LPCVD grown graphene for ultrasensitive detection of carbaryl*, Sens. Actuators B,**260**, pp.78-85,(2018).
9. D. Chan, T.W. Healy, L.R. White, *Electrical double layer interactions under regulation by surface ionization equilibria-dissimilar amphoteric surfaces* ,J. Chem. Soc. Faraday Trans. 1: Phys. Chem. Condens. Phases, **72**, pp. 2844-2865 (1976).
10. L. Bousse, N.F. De Rood , P. Bergveld, *Operation of chemically sensitive field-effect sensors as a function of the insulator-electrolyte interface* . IEEE Trans. Electron. Dev., **30**,10, pp. 1263-1270(1983).
11. N. Chaniotakis, *Biological Identification // Bio-chem-FETs: field effect transistors for biological sensing*,(2014).
12. Y. Cui, Q. Wei, H. Park, C.M. Lieber, *Nanowire nanosenosrs for highly sensitive and selective detection of biological and chemical species*, Science ,**293**, pp.1289–1292(2001).

Using Field-Effect Gas Sensors for Monitoring H_2 in Transformer Oil

Artur Litvinov[1], *Nikolay* Samotaev [1,*], *Maya* Etrekova[1], *Anastasia* Ivanova[1], *Dmitriy* Filipchuk[1] and *Yuri* Klishin[1]

[1] National Research Nuclear University MEPhI (Moscow Engineering Physics Institute), Microelectronics department, 115409 Moscow, Russian Federation

Abstract. Hydrogen can be released during the thermal decomposition of organic materials, therefore, monitoring its level in the working industrial high-voltage transformer oil allows you to identify the development of degenerative processes in advance, because these processes can lead to an accident in the future. In experiments has shown that highly sensitive and small-sized field effect gas sensor based on the metal-insulator-semiconductor structure can be used for measuring of Hydrogen in oil with direct contact of its structure with transformer oil. Given the harsh environmental conditions of hydrogen measurement the field effect capacity type gas sensor were fabricated by using laser micromilling technique for fabrication compact ceramic surface mounting device package and microheater for sustentation working temperature of metal-insulator-semiconductor structure.

1 Introduction

Transformers are one of the main components of the complex equipment of many energy (power plants, substations, converting devices) or industrial enterprises. In order to avoid equipment failure, it is necessary to conduct transformer oil tests in a timely manner. More precisely, a check is made of its quality. Periodic monitoring of transformer oil is one of the components of technological maintenance of industrial equipment in enterprises. The main characteristics of transformer oil, its purity and useful properties determine the performance of transformers. The ability of the oil to maintain its original properties in operating equipment during operation is called the stability of transformer oil. If the power equipment has no defects and works without failures, then the characteristics of the new oil practically do not change. Fresh transformer oil has a light color and certain conforms to certain standards that determine its dielectric and physico-chemical properties. During operation, the stability of transformer oil is significantly reduced, noticeable changes in performance appear and the oil darkens.

The quality tests determine the start of the aging process of transformer oil. Based on the tests of transformer oil, its performance is assessed, the necessary cleaning and recovery procedures are prepared, and comprehensive reports on the overall operational status of the transformer equipment are prepared [1]. In transformer oil, it undergoes a thorough analysis due to the fact that water and air are one of the main catalysts for the aging process of working fluids. Moisture content is measured by the amount of hydrogen in the interaction of transformer oil with calcium hydride for a specified time [2]. The gas content is calculated using an absorbiometer [3], Fourier transform infrared spectroscopy [4] or chromatograph [5].

Hydrogen can be released during the thermal decomposition of organic materials; therefore, monitoring its level in the working industrial high-voltage transformer oil allows you to identify the development of degenerative processes in advance, because these processes can lead to an accident in the future. In experiments has shown that highly sensitive and small-sized field effect gas sensor based on the metal-insulator-semiconductor (MIS) structure can be used for measuring of H_2 in oil with direct contact of its structure with transformer oil. Numerical estimates show that with the diffusion coefficient D = 4*10-9 m^2/s [6] (at room temperature), the time required for H_2 saturation in a transformer oil with 10 cm depth to a level $0{,}9C_0$ is more than 1 year. In reality, there are convective flows occurred due to the temperature gradient which are combined with diffusion. Nevertheless, such a large diffusion time must be taken into account when designing the experimental setup and the obtained results interpreting.

2 Experimental

A schematic representation of the field-effect gas sensor is shown on Figure 1 and described in work [7]. The measuring H_2 concentration with the field-effect gas sensor is as follows. When hydrogen molecules from the external environment interact with the metal electrode, H_2 molecules decompose into atoms and diffuse to the

* Corresponding author: nnsamotaev@mephi.ru

metal-insulator interface. It is known that on the surface of palladium hydrogen atoms have a dipole moment. The electric field of dipoles leads to a change in the electric field in the dielectric and in the surface layer of the semiconductor. As a result, the capacitance of the MIS-structure changes, the value of which is fixed by the electronic unit while maintaining a constant bias voltage on the MIS-structure.

Given the harsh environmental conditions of hydrogen measurement the field effect capacity type gas sensor were fabricated by using laser micromilling technique for fabrication compact ceramic surface mounting device package and microheater for sustentation working temperature of MIS structure. The technology of fast fabrication functional sensor in specific ceramic SMD package with base 9x9 mm was precisely described in work [8] and photo of sensor package is shown on Fig. 1. A significant problem remains the difference in the energy consumption of sensors located in the air and immersed in transformer oil due to the difference in the thermal conductivity of mediums. The use of ceramics for the manufacture of the sensor package not only reduces the overall size of the sensor, but also allows to reduce this dependence on the thermal conductivity of the ambient medias in comparison with the use of standard metal casings for gas sensors used by us in other our works where it is more important for selectivity and decrease time of sensor response [9-10].

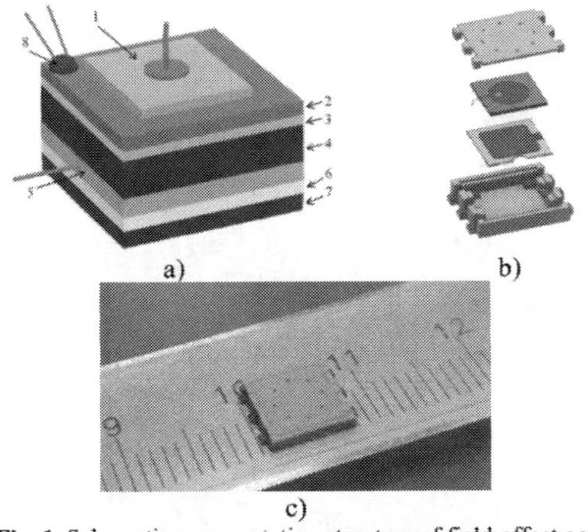

Fig. 1. Schematic representation structure of field-effect gas sensor in SMD package. (a) MIS-structure, which is a capacitor, the substrate of which is an n-type silicon wafer (4), 3 - coated on one side with a SiO_2 film and 2 - an additional layer of Ta_2O_5; 1 - catalytic metal Pd film is applied to the dielectric by laser sputtering, to which an external metal electrode is glued; 5 - the second electrode is glued to the metal layer deposited on the back of the capacitor (semiconductor); 6 - the MIS-structure is located on the insulating plate, 7 - separating it from the film resistive heater. 8 - the temperature of the sensitive element is measured and adjusted with a thermistor. (b) 3D model of package and microheater in *.stl format (screenshot in the Autodesk Inventor CAD program) before ceramic package fabrication (c) External view already assembled MIS sensor in SMD package (dimensions scale are in mm).

The sketch of setup used in experiment imitates similar to a transformer conditions of H_2 diffusion present on Fig. 2. Both field-effect sensors were preliminarily calibrated for H_2 in air (in the concentration range 5-100 ppm) at the sensing element temperatures of 23 and 100°C. The optimal temperature of the sensor structure is 100-150°C. The total capacity of the tank used in experiments is 1 liter, and the volume of transformer oil used is 0.1 liter. The height of the oil column is 2 cm.

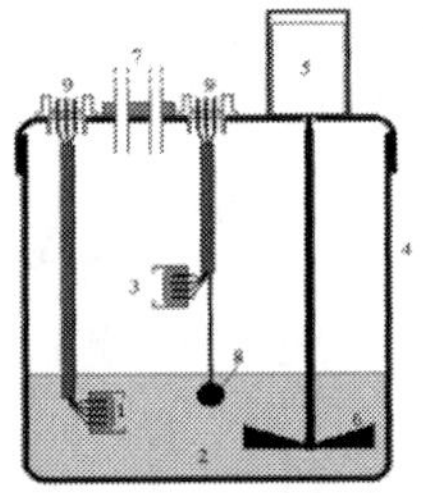

Fig. 2. Schematic representation of the experimental setup: 1 - the main field-effect sensor; 2 - the structure of the sensitive element of which interacts with transformer oil; 3 - the reference field-effect sensor, located in the air gap above the oil, are placed in 4 - an airtight container; 5 - the electric motor rotates for 6 - the mixing device; 7 - the tubes are used to supply the specified concentration of hydrogen and ventilation of the container; 8 - the temperature of the oil is controlled with a thermocouple; 9 - vacuum electrical connectors are used for electrical input.

In Table 1 measurement results of the field-effect sensor sensitivity to hydrogen in air and in transformer oil are presented. S – concentration sensitivity H_2 5 ppm; A_n– signal noise amplitude ;$\tau_{0,9}$ and $\tau_{0,1}$ signal saturation time to a level of 0.9 at H_2 concentration of 5 ppm and signal relaxation time to a level of 0.1 from a signal recorded at 100 ppm; T_{sensor} – sensor working temperature.

Table 1. Measurement results of the field-effect sensor sensitivity to hydrogen in air and in transformer oil.

Parameter	In air	In oil	T_{sensor}, °C
S, pF/ppm	5,2	2	23
A_n, ppm	0,4	0,8	
$\tau_{0,9}$, hour	5,5	5,9	
$\tau_{0,1}$, hour	10	>12	
S, pF/ppm	5,5	-	100
A_n, ppm	0,3	-	
$\tau_{0,9}$, min	8	-	
$\tau_{0,1}$, min	17	-	

3 Results and Discussion

During experiments, we have target the using main sensor under ambient temperature for excluding excess power consumption for maintaining working temperature in intense heat exchange with oil. For this reason, its heating element was turned off and equal to the temperature of the oil and was 23°C. The reference sensor was maintained at a temperature of 100°C. Relative humidity above oil was 25%. The experiment was carried out as follows. Through drainage tubes, fixed H_2 concentrations in the range from 5 ppm to 100 ppm were fed into the container. After the mixture was supplied, the tubes were blocked and, subsequently, the H_2 concentration in the vessel was monitored by sensor (Fig. 2, position 3).

To accelerate the process of oil saturation with H_2, a mixing device was used, the rotation frequency of which was selected to exclude the formation of air bubbles in the oil. Figure 3 shows the dependence of the response of the sensors (main and reference) on the concentration of H_2. Table 1 shows the measurement results.

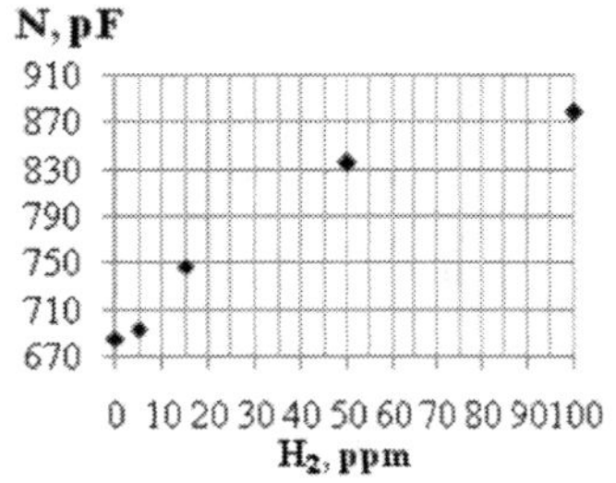

a)

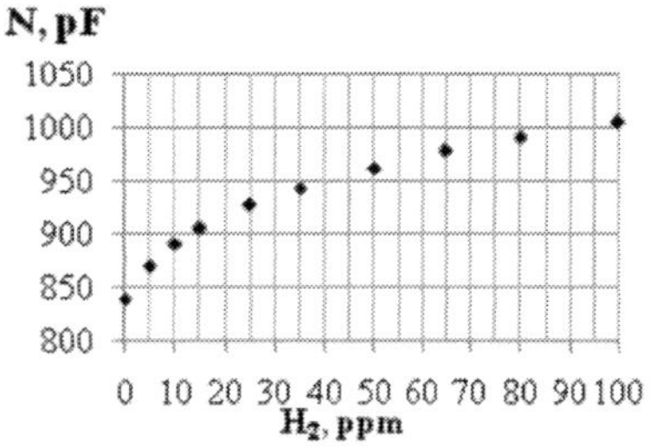

b)

Fig. 3. The static characteristics of the field-effect sensor are: (a) the structure of the sensor (working temperature 23°C) is in contact with transformer oil, (b) the structure of the sensor (working temperature 100°C) is in contact with air.

The experiment showed that the temperature of the sensor has a significant effect on the detection time of H_2. Immersion of the sensor in oil leads to an increase in the noise level of the signal and a decrease in the sensitivity of the sensor in the low concentrations range. Also, the presence of transformer oil does not change the functional dependence of the sensor capacity on the H_2 concentration. The response time to H_2 is significantly less than the time of its diffusion in oil, which makes it possible to use the sensor as a sensitive element in the control systems for the level of dissolved H_2 in the oil of a working transformer under ambient temperature.

4 Summary

It has been demonstrated that by using laser micromilling technology possible to fabricate custom integrated microheater platform with SMD package for MIS gas sensor. The design is based on monolithic ceramics, the use of platinum as the material of the heater and the thermometer creates a mechanically robust design that is resistant to aggressive transformer oil media. Practical experiments have shown that for the sensitivity of the sensor, the difference in the composition of the gaseous medium and hydrogen dissolved in oil is negligible. From a physical point of view, the difference in energy consumption and operating temperature is manifested in the response rate to hydrogen, which differs by 35 times for air and oil, respectively. Since the rate of increase in the concentration of hydrogen in the transformer oil relative to the sensor response time is small, it is preferable to use the sensor at ambient temperature without exposing it to unnecessary temperature stresses and unreasonably excessive energy consumption. The developed design of field-effect sensor can be used as an inexpensive tool for on-line monitoring quality of transformer oil, without resorting to alternative expensive analyzes.

Acknowledgment

This research was funded by the Israel Innovation Authority, BMBF (Federal Ministry of Education and Research) in Germany with funding No. 02P15B520, and the Ministry of Science and Higher Education of the Russian Federation funding with unique identifier RFMEFI58718X0054 in frame of MANUNET project MNET17/ADMA-1147.

References

1. W.H. Tang, Q.H. Wu, Springer, London (2011)
2. V. V. Kondalkar, G. Ryu, Y. Lee, K. Lee, Sensors and Actuators B: Chemical, **286** 377-385 (2019)
3. Y. Gui, W. Li, X. He, Z. Ding, L. Xu, Applied Surface Science, **507** article 145163 (2020)
4. J. Jiang, Z. Wang, G. Ma, H. Song, C. Zhang, Optik, **176** 214-220 (2019)
5. Y. Leblanc, R. Gilbert, M. Duval, J. Hubert, Journal of Chromatography A, **633** 185-193 (1993)
6. A.L. Bychkov, S.M. Korobeynikov, A.Y. Ryzhkina, Technical Physics, **56** (2011) 421-422
7. I.N. Nikolaev, L.N. Kalinina, A.V. Litvinov, Automation and Remote Control, **72(2)** 442-448 (2011)
8. N. Samotaev, A. Litvinov, M. Etrekova, K. Oblov, D. Filipchuk, A. Mikhailov, Sensors, **20(5)** article 1514 (2020)
9. B. Podlepetsky, M. Nikiforova, A. Kovalenko, Procedia Engineering, **168** 251-254 (2016)
10. B. Podlepetsky, N. Samotaev, M. Nikiforova A. Kovalenko, Sensors, **19(8)** article 1855 (2019)

Research and Application of a Data Processing Method on Outliers in Unmanned Aerial Vehicle (UAV) Tracking Measurement

Kunkun Li*, *Rui* Cao, *Xiaoyu* Qi, and *Yaxiong* Hu

Northwest Institute of Nuclear Technology, China

Abstract. In unmanned aerial vehicle (UAV) flight test, some complex outliers often appear in tracking measurements due to various factors stemming from environment, instrumentations, and even operators. This seriously affects the reliability of applications and analysis. This paper aims at the problem of outliers elimination to measurement data and proposes a joint data processing method comprising grouped data extraction, data trend modeling, and outliers detection. In the context of accuracy assessment test for measurement and control equipment (MCE), an approach using the measurement parameter for UAV flight is designed to estimate the time system deviation between MCEs, as well as to validate the effect of the proposed joint method. Finally, a simulation is implemented and results show the effectiveness and feasibility of the joint method to outliers elimination and time deviation estimation.

1 Introduction

With the development of advanced technology throughout the world, *Aircraft flight test* (AFT), an indispensable technique in tasks such as aircraft design assistance, quality inspection, and performance appraisal, plays an increasingly important role in the coming into service and capabilities upgrade of aircrafts [1]. As an unusual branch *unmanned aerial vehicle* (UAV) flight continually attracts the interests of scholars from various research domains owing to its widespread availability and immense demand. In such a flight test, often times the tracking measurements serve as a basis of assessment for UAV flight quality, and tracking performance of ground-based *measurement and control equipment* (MCE). Hence, the measurement data quality directly affects the reliability of following applications and analysis.

Regrettably, during data acquisition some data points whose patterns do not conform to expected normal behavior of measurement data often occur in sequential tracking measurement data. These anomalous patterns or abnormal data points are referred to as outliers by analyst [2]. Concerning the causes to outliers, they typically come from internal impact, vibration, system failure from within instrumentations, environmental disturbance, and even some inappropriate operations by people. To separate the useful information from measurement data, it is, therefore, necessary to do some work including outliers detection, elimination, and correction in the preprocessing phase.

Generally, for a simple one-dimensional data set in time series, the outliers in it can be classified into two basic types according to their continuity, namely isolated outliers and spot outliers, respectively. As they are called, the isolated outliers refer to some discontinuous abnormal data points since accidental disturbances arise at times. And the spot outliers refer to some continuous abnormal data points generally with little difference in amplitude due to some persistent-existing causes [3]. In previous works [4-7], the outlier detection approaches towards isolated outliers revolve around some conventional statistical detection methods. They have well elimination performance but are not always tolerant to spot outliers owing to limited recognition capacity, and constraint on the statistical property of data. Thus they suffer from problems of halfway outliers elimination, and even removal of some useful information. As for spot outliers, the techniques mainly include Kalman filtering method, robust filtering with M-estimation method, B-spline approximation by least square method, etc. These methods offer better recognition to spot outliers, whereas they still pose challenges such as erratic processing performance, and data loss after outliers elimination.

The remainder of this paper is organized as follows: First, a joint data processing method on outliers is formulated, where the Pauta criterion is introduced and the involved model or thought are presented. Then, a problem concerning the estimation of time system deviation between MCEs in accuracy assessment test is described. Further, a solution using the tracking measurements of UAV flight is designed, and next the estimation steps are elaborated in details. After that a simulation on an UAV flight test scenario is executed, and the performance of some estimation methods are compared and analysed. Finally, this paper summarizes the main contributions and draws a conclusion.

* Corresponding author: lkk9419nudt@163.com

2 Joint data processing method

2.1 Pauta criterion

The Pauta criterion [8], a typical and practical method in data preprocessing, is widely used for scientific computation on some engineering problems. According to the criterion, for a long series of sample data points which follow a normal distribution, let u denote the mean value and σ be the standard deviation. Theoretically, u is 0 in terms of the residuals between u and the sample data values, and the probability of residuals within the confidence interval $(u-3\sigma, u+3\sigma)$ is specified over 99.73%. Hence, some of the sample data are regarded as outliers while the corresponding residuals exceeds such interval with a less 0.3% probability.

Consider a sequence of measurement data points denoted by $x(k)\ \{k=1,2,...,N\}$, where N represents the data length. Suppose some outliers are present at uncertain positions in data series. To detect outliers via Pauta criterion, the arithmetic mean value of measurement data, denoted by $\bar{x}$, needs to be calculated as follows

$$\bar{x}=\frac{1}{N}\sum_{k=1}^{N}x(k) \tag{1}$$

and the standard deviation σ is given by

$$\sigma=\sqrt{\sum_{k=1}^{N}\frac{(x(k)-\bar{x})^2}{N-1}} \tag{2}$$

With $\bar{x}$, the residuals of measurement data can be expressed by

$$x_r(k)=x(\text{k})-\bar{x},\ k=1,2,...,N \tag{3}$$

Compare the absolute value of residuals above with a threshold 3σ in sequence, and conduct the critical condition. Once $|x_r(k)|\geq 3\sigma$ holds, the data at point k, viz., $x(\text{k})$, is then perceived as an untrusted data or outlier, and thus should be eliminated, otherwise be preserved.

2.2 Data trend modeling by least square

One of the key to data analysis lies in a good mathematical model chosen to enhance the extraction of useful information. However, a congruent model is hardly established in most cases owing to lack of priori information, and limited data collection. With such challenges, numerous methods have been developed and applied for data feature extraction over the past several decades [9-11], and amongst them the *least square* (LS) method is tractable to analyst from various knowledge disciplines due to its simplicity, robustness, and convenience of implementation.

The steps of data trend modeling with least square method are

Step 1. Suppose an approximate polynomial with a certain degree is exploited to data trend fitting. List a matrix equation according to the LS theorem;

Step 2. Solve the equation to yield a coefficient matrix of the above polynomial, and then we derive a curve fitting data trend.

To be more specific, let $x(t_n)\ \{n=1,2,3,...,N\}$ denote the time-series data and t_n be the discrete sampling time. Then we employ a Kth-degree polynomial to fit the data trend, as expressed by

$$y(n)=\sum_{k=1}^{K}b_k\left(t_n\right)^k,\ n=1,2,...,N \tag{4}$$

where b_k represents the polynomial coefficient matrix. In other words, $y(n)$ is taken to be the approximation of polynomial elements w.r.t. $x(n)$. Under such condition an error function is defined as

$$\begin{aligned}E(h)&=\sum_{n=1}^{N}(x(n)-y(n))^2\\&=\sum_{n=1}^{N}\left(x(n)-\sum_{k=1}^{K}b_k\left(t_n\right)^k\right)^2,\ n=1,2,...,N\end{aligned} \tag{5}$$

To find the minimum of $E(h)$, let us take the partial derivative of equation (5) w.r.t. b_j and make it to be zero, then we have

$$\frac{\partial E}{\partial b_j}=\sum_{k=1}^{N}2\left[x(n)-\sum_{k=1}^{K}b_k\left(t_n\right)^k\right]\left[-\left(t_n\right)^j\right]=0,\ n=1,2,...,N \tag{6}$$

After algebraic transposition, $K+1$ equations are generated as follows

$$\sum_{k=0}^{K}b_k\sum_{n=1}^{N}\left(t_n\right)^{k+j}=\sum_{n=1}^{N}x(n)\left(t_n\right)^j,\ n=1,2,...,N;\ j=0,1,...,K \tag{7}$$

where b_k could be got by means of matrix inversion. Ultimately, a resulting polynomial for the data trend is obtained.

2.3 Outliers elimination by grouped data extraction

In the research field of engineering, some knotty problems with large scale computing are likely encountered. Most often, some straight-forward solutions would probably offer the corresponding answers in a general sense. Nevertheless, this do also face a significant drawback that such doing not merely wastes researchers' energy but imposes great burden on the computer's processor in attempt to support fast and accurate calculation. With this challenge, some of the problems are sometimes split into the same but smaller

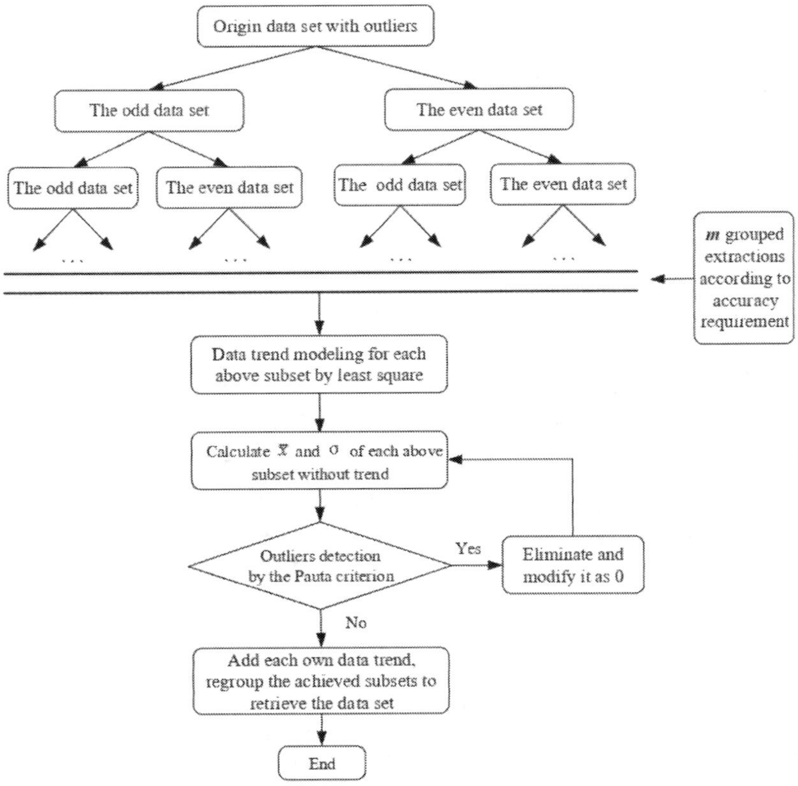

Fig. 1. Flow diagram of outliers elimination by the joint data processing method.

sub-problems by researchers according to the divide and conquer method [12]. As a consequence, the answers of these sub-problems are relatively easier to get, and after a comprehensive analysis, the resolution of original problem is achieved in the end.

With this flexibility in mind, a time-series data set with complicated types of outliers can be decomposed into multiple subsets with smaller sizes. For example, one sequential data set becomes 2^m subsets after m recursive grouped extractions according to the differences of odd and even positions of data points. This way there are two advantages to data processing, as follows

(1) Spot outliers lead to a problem that the calculated σ of original data deviates from the actual σ to some extent. After multiple grouped extractions the number of outliers in each resulting subset decreases, and thus the σ of such a subset tends to the actual σ ;

(2) The types of outliers in original data are simplified as all spot outliers resolve into some sparse isolated outliers after several grouped extractions. Thus it is convenient for processing of each subset.

Given all the above argument, for elimination of outliers in a sequential data set we can handle it with a considered joint method on the basis of grouped data extraction, data trend modeling, and outliers detection via Pauta criterion, as shown in Fig. 1.

3 Method application

3.1 Problem description

In general, the *measurement and control equipment* (MCE) consists mainly of some radio or optical measurement equipment (radar, photoelectric theodolite), and some wireless information transmission equipment (telemetry equipment, remote control equipment). Before delivery of a MCE, it is of significance to assess its indicators in a testing outfield with some dynamic conditions in consideration, that is, MCE accuracy assessment test [13]. Such a test relates to ceaseless calibrations for the measurement error of MCE, and most importantly, obtains a comprehensive evaluation of MCE's performance. As such, the capability of MCE is ensured to meet the functional index listed in technical specifications, and customer's requirements.

Nowadays, with the advances of high-precision carrier phase differential GNSS (Global Navigation Satellite System), a GNSS equipment with centimeter-level positional accuracy is generally mounted on the UAV platform as a cooperative target in MCE accuracy assessment test by testers, as illustrated in Fig. 2. The raw information on time-variant position of UAV flight is then transmitted to a receiver aboveground in real time

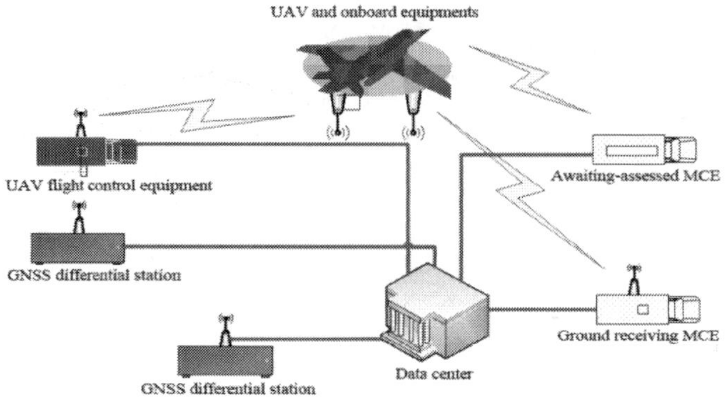

Fig. 2. Schematic diagram of some participant equipments in accuracy assessment test.

via a S-band telemetry transmitter. Finally, high-precision differential positioning information is achieved in ground data center, and acts as a comparison standard with regard to the MCE tracking measurements. Thereby, a statistical analysis of the comparison residuals can be conducted so as to accomplish MCE accuracy assessment.

Suppose a tracking radar is the awaiting-assessed MCE without loss of generality. And some other aided testing MCEs are assumed to have been all tested aforehand to reach their ideal state in test. Before starting a comparison analysis, the radar tracking measurement data, along with the standard data for comparison, must be in a unified time system as far as possible. However, there are several potential causes for disunity of time systems of different MCEs, as follows

(1) The precision of relevant time systems unification is finite in the strict sense. Even if the timing accuracy of the satellite positioning system is nominally up to a nanosecond level, it may appear that the actual time system deviation reaches tens of milliseconds since some differences always exist in signal processing, and integration of terminal timing systems for different manufacturers;

(2) Not all of MCEs, in practice, adopt the same strategy to record time information on measurement data. Some mark time on the original recorded data, while others add time stamps to the post-processing data;

(3) Uncertain time delays in data processing, transmission, and other links may cause the ideal time synchronization not to be achieved even in the case of ideal time system.

Hence, it is of course significant to obtain a well estimate of time deviation between the radar system and other ideal MCE system, and unsatisfactory estimation will directly affect the final success of accuracy assessment test.

In short, what problems to address in this paper are

Problem 1. *How to restrain the interference of complex outliers in radar tracking measurements so as to ensure the reliability of applications?*

Problem 2. *How to obtain a well estimate of the time deviation so as to conduct the comparison analysis under an aligned time system?*

3.2 Problem solution

In this part, the two preceding problems which seem uncorrelated are considered together in the context of accuracy assessment test. And an approach using the pitch angles of UAV flight relative to the tracking radar is designed aimed at the estimation of time system deviation while regarding the elimination of outliers in tracking measurements.

In order that the standard data mentioned earlier has a stable and reliable quality for comparison, the UAV route, in accuracy assessment test, is generally designed in form of a horizontal straight line, and UAV flies at a constant speed to stay in a cruise state. Additionally, it remains to require that the tracking capability of the awaiting-assessed radar must be taken into account when we perform planning operation on the route. Subsequently, according to a feature that a turning point of the tracking parameter emerges as the flying-along-route UAV passes a specific waypoint, whose projection on ground is closest to the radar, we get an idea of obtaining the time deviation estimation.

As shown in Fig. 3, the position of radar is denoted by P_o, P_c represents the aforementioned waypoint in route. D is the distance between P_o and the projective point of P_c. θ is the theoretical angle of UAV relative to radar in pitch at time t. h is the flight height of UAV relative to the ground. Let t_0 denote the starting time of UAV entering the straight-line route and t_c be the time of UAV reaching the waypoint P_c. At this point θ can

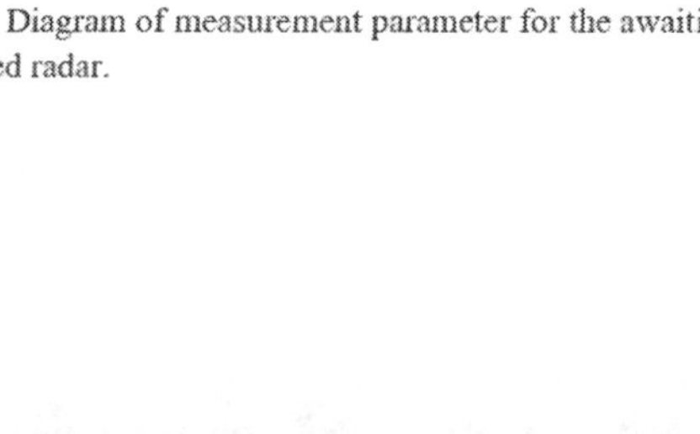

Fig. 3. Diagram of measurement parameter for the awaiting-assessed radar.

be expressed in the following mathematical formula

$$\sin\theta(t) = \frac{h}{\sqrt{v^2 \cdot (t+t_0-t_c)^2 + D^2 + h^2}} \tag{8}$$

then we have

$$\sin^{-2}\theta(t) = \frac{v^2 \cdot (t+t_0-t_c)^2}{h^2} + (1+\frac{D^2}{h^2}) \tag{9}$$

As can be seen from formula (9), $\sin^{-2}\theta(t)$ is explicitly characterized by a quadratic polynomial with respect to the parameter t. This heuristically allows us to make data trend modeling by least square method with $K=2$. Let us refer to $\sin^{-2}\theta_1(t_1)$ and $\sin^{-2}\theta_2(t_2)$ as another two transformed forms of the standard data for comparison, and radar tracking measurements for assessment, respectively. In the presence of time system deviation the two corresponding theoretical curves are indicated in Fig. 4, where the extreme points on two curves uniformly stand for a specific time at which UAV reaches the waypoint P_c.

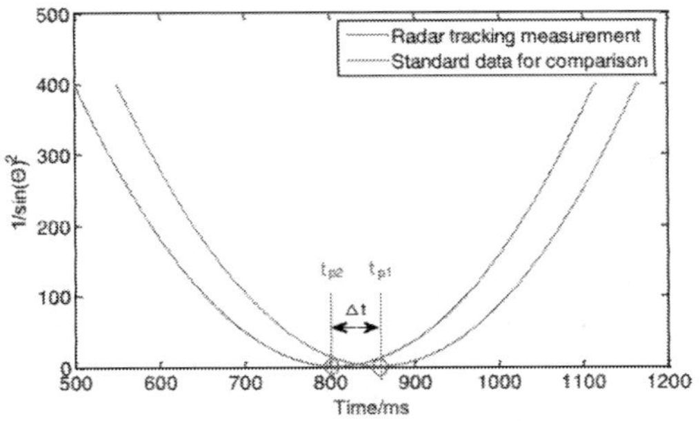

Fig. 4. Theoretical curves of the two $sin^{-2}\theta(t)$ in presence of time deviation.

Following this invariant uniformity, it suffices to consider the deviation between the two associated times of such extreme points, which can be readily calculated by the following steps

Step 1. Use two quadratic polynomial curves to fit $\sin^{-2}\theta_1(t_1)$ and $\sin^{-2}\theta_2(t_2)$, respectively;

Step 2. Find approximate times for extreme points of the two curves above, denoted by t_{p1} and t_{p2}, respectively;

Step 3. Calculate the time system deviation by $\Delta t = t_{p1} - t_{p2}$.

3.3 Solution flow

To get the standard data for comparison, the high-precision GPS information on position of UAV flight is obtained by the real-time kinematic (RTK) positioning technique in data center, and converted to the pseudo pitch angles of the tracking radar after a coordinate transformation. This mathematically results in a standard data sequence, denoted by $E_{S\tan}(t_k)$ $(k = 1,2,3,...,N_{S\tan})$ in discrete form for brevity. In parallel, the radar tracking measurements in pitch are denoted by $E_{Rad}(t_n)$ $(n = 1,2,3,...,N_{Rad})$. With some underlying disturbances inside and outside, $E_{Rad}(t_n)$ can be expressed in the following form

$$E_{Rad}(t_n) = E_{Rad}(t_n)^* + s_n + \omega_n + \delta_n \tag{10}$$

where $E_{Rad}(t_n)^*$ represents the actual pitch angle sequence; s_n is the systematic error which can be removed by operations such as modeling, compensation, and calibration; ω_n is the random error which can be restrained by filtering, and smoothing methods; δ_n signifies some complicated types of outliers, and the outliers elimination, as one of the two problems stated previously, is concerned in this paper.

Recalling the content in Problem solution, the estimate of time system deviation can be obtained analytically with the proposed joint method, as steps presented below

Step 1. Conduct data trend modeling for $\sin^{-2}E_{S\tan}(t_k)$ by least square method. We derive a fitting quadratic polynomial $f_{S\tan}(t_k) = a_1 t_k^2 + a_2 t_k + a_3$. Then the time corresponding to the extreme point is $t^P_{S\tan} = -a_2/2a_1$;

Step 2. Group $\sin^{-2}E_{Rad}(t_n)$ into 2^m data sequences via m recursive extractions, as depicted in Fig. 1;

Step 3. Conduct data trend modeling for each above data sequence by least square method, and subtract each own trend polynomial. We get 2^m data sequences without data trends.

Step 4. Use Pauta criterion to detect the resulting 2^m data sequences above. Eliminate the discovered outliers and correct them as 0;

Step 5. Add each own data trend for the newly resulting 2^m data sequences above and regroup them. We get a new data sequence $\sin^{-2}\hat{E}_{Rad}(t_n)$. Conduct data trend modeling for it by least square method. We derive another fitting quadratic polynomial $f_{S\tan}(t_n) = b_1 t_n^2 + b_2 t_n + b_3$. Then the time corresponding to the extreme point is $t^P_{Rad} = -b_2/2b_1$;

Step 6. Calculate the time system deviation by $\Delta t = t^P_{S\tan} - t^P_{Rad}$.

3.4 Simulation and analysis

To verify the effectiveness of the proposed method applied for estimation of time system deviation, a simulation scenario for accuracy assessment test is created and carried out. Suppose the test for tracking radar is executed on a virtual testing outfield. Some other MCEs participate in the test as aided testing equipments in different locations, which are not displayed for simplicity.

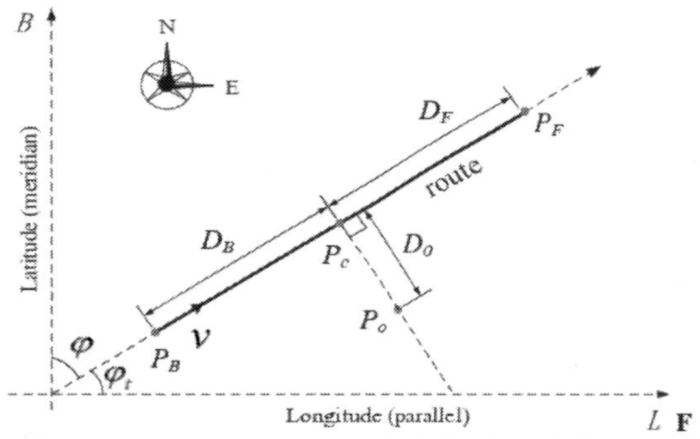

Fig. 5. Accuracy assessment test in the horizontal plane (P_B—initial route point, P_F—final route point, P_O—radar location).

For the planning of UAV route, Fig. 5 shows its schematic diagram in the horizontal plane. At 12:01 PM, UAV flies along the planned route P_BP_F at a constant cruising speed $V = 50m/s$ from southwest to northeast on the side of radar. The flight height is $h = 1500m$. The flight angle is $\varphi = 45^{\circ}$. The length of route is $P_BP_F = 10km$, and herein we postulate that the projective point P_c is selected as half of the route (i.e., $D_B = D_F = 5km$). The radar, denoted as P_o in Fig. 5, is located a distance of $D_o = 5300m$ away from the route.

For the awaiting-assessed radar, suppose the measurement noise listed in technical specification for angular measurements is zero-mean white Gaussian noise with random walks of $\sigma_E^2 = (300\mu)^2\text{rad}^2$ in pitch, and $\sigma_A^2 = (300\mu)^2\text{rad}^2$ in azimuth, respectively. The sampling frequency is $f_s = 20\text{Hz}$. As for the GPS information on UAV position generated by the real-time kinematic (RTK) positioning technique in ground data center, the positional accuracy along three axes in the earth-centered earth-fixed coordinate system are assumed to be all equal to $0.01m$.

Finally, suppose the radar and ground data center work under a nominally unified GPS time reference, but, owing to some of the causes explained previously, there exists a time deviation between them, which here is assumed as $\Delta t = 35\text{ms}$. To ensure that the tracking measurement data for comparison analysis really comes from a data collection regarding the straight UAV flight phase, we selectively consider a 7km range in route (i.e., a length of 3.5km for the front, and back sub-routes, respectively) since UAV usually calls for adjustment on its own attitude to fly into and off a route at the start and end phase, respectively. With some various disturbances, there are some complex outliers in measurements, and they appear more intensively than others as UAV approaches P_c. This is mainly due to instable tracking quality induced by the increasing azimuth velocity to reach its limit over time.

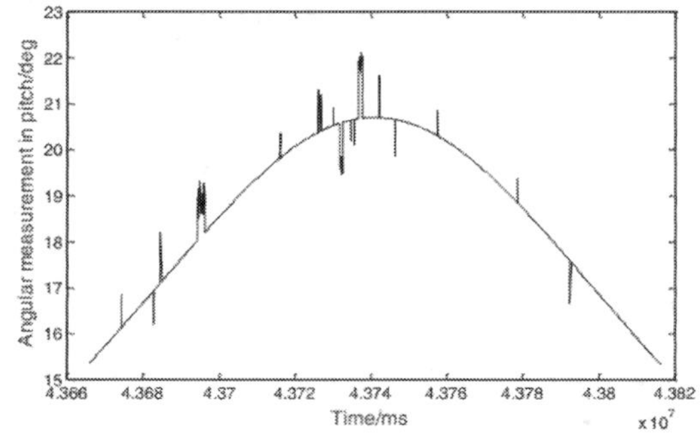

Fig. 6. Curve of radar tracking measurement in pitch.

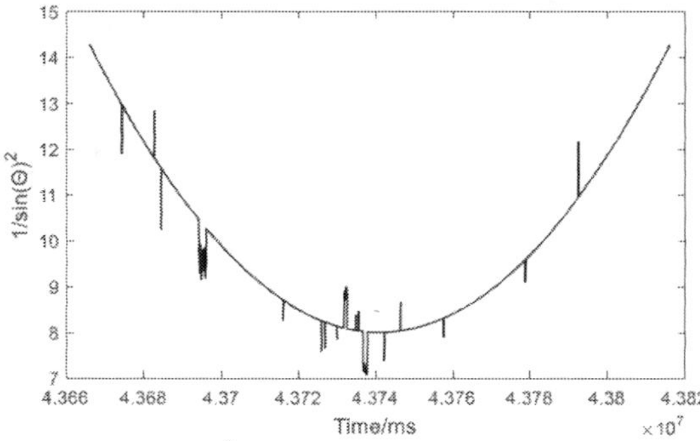

Fig. 7. Curve of $sin^{-2}\theta(t)$ by converting the form of radar tracking measurement in pitch.

Figs. 6 shows the changing curve of radar tracking angular measurement in pitch. It can be seen that the curve has a distinct turning, but, in the strict sense, it has no obvious feature of polynomial with an explicit degree. In this case, we convert the form of pitch angle measurements to what as formula (9), and then a curve with the feature of quadratic polynomial for $\sin^{-2} E_{Rad}(t_n)$ is derived, as depicted in Fig. 7.

Figs. 8, 9, and 10 show the curves of $\sin^{-2} \hat{E}_{Rad}(t_n)$ after conducting $m = 1$, 2 and 3 recursive grouped extractions, and outliers elimination, respectively. A quick look at the three Figs reveals that the number of outliers decreases as m increases, thereby showing the effectiveness of the proposed method to outliers elimination. For a clear and fair comparison on the estimation effect, we make 200 independent Monte Carlo runs. Table 1 provides the average estimates of time system deviation by different methods. From Table 1 it is apparent that the estimate, after more recursive grouped extractions, comes closer to the actual value we

Table 1. A comparison of average estimates of time system deviation by different methods.

Methods	Straight-forward polynomial curve fitting method	Joint data processing method by $m = 1$	Joint data processing method by $m = 2$	Joint data processing method by $m = 3$
Average estimates	421.28ms	46.34ms	40.60ms	37.16ms

Deviation from $\triangle t$=35ms we set	386.28ms	11.34ms	5.60ms	2.16ms

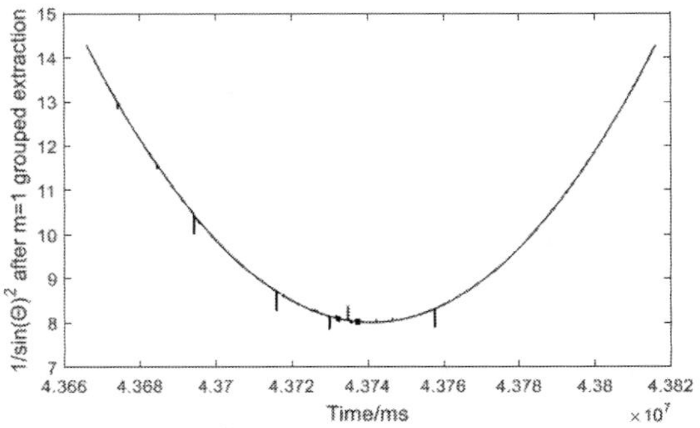

Fig. 8. Curve of $sin^{-2}\theta(t)$ after m=1 grouped extraction and outliers elimination.

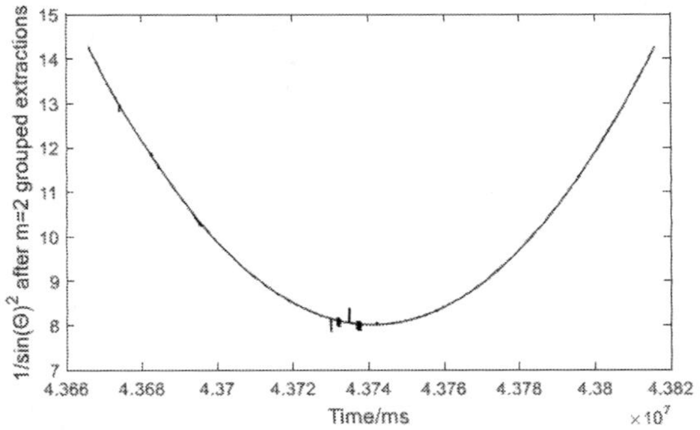

Fig. 9. Curve of $sin^{-2}\theta(t)$ after m=2 grouped extractions and outliers elimination.

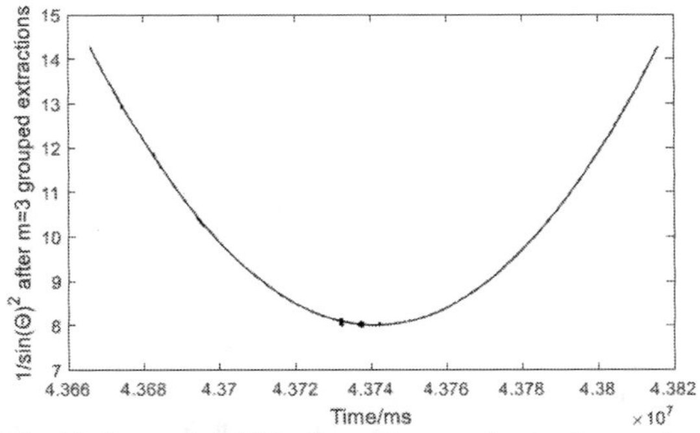

Fig. 10. Curve of $sin^{-2}\theta(t)$ after m=3 grouped extractions and outliers elimination.

set previously. Moreover, the joint method performs noticeably better compared to a straight-forward polynomial curve fitting method, whose estimate being considerably far away from the actual value indicates the adverse impact of complex outliers in case of ignoring outliers elimination. Consequently, the proposed method is feasible to the estimation of time system deviation, and thus helpful to the accuracy assessment test.

4 Conclusion

A joint data processing method, constituted by grouped data extraction, data trend modeling, and outliers detection, was presented to eliminate complex outliers in UAV tracking measurements, and sought to estimate the time system deviation between different measurement and control equipments. A simulation on accuracy assessment test for radar was implemented, and results demonstrated that the proposed joint method was effective to outliers elimination. What's more, the method helped obtain a well estimate of time system deviation, thereby facilitating the accuracy assessment test for the awaiting-assessed equipment.

References

1. M. Zieja, W. Kosiński, J. Wójcik, J. Kon, **20**, 501-510 (2013)
2. V. Chandola, V. Kumar, ACM. Comp. Surv, **41**, 3 (2009)
3. S.H. Cui, M. Wang, J. Wang, *Velocity Tracking Measurement Data Processing* (NDI Press, Beijing, 2017)
4. B.W. Hou, J.Q. Wang, X,Y. Zhou, D. Li, Z.M. He, Aero. SH, **35**, 91-100 (2018)
5. H.B. Zhou, D.D. Chen, Y. Peng, B, Shipb. Elec. Coun, **36**, 84-87 (2013)
6. Z. Liu, J.F. Wang, Y. Wu, *Proceeding of the 32th Chinese Control Conference*, 4652-4655 (2013)
7. Y.P. Duan, EMT, **35**, 40-42 (2012)
8. D. Xiao, B. Shang, B. Wu, *International Conference on Fuzzy Systems & Knowledge Discovery* (2015)
9. P. Theo, *Algorithms for Graphics and Image Processing* (CS Press, 1982)
10. M. Fischler, *Commun Acm*, **24** (1981)
11. W. Zeng, S. Feng, Inf. Sci, **272**, 73-83 (2014)
12. M. Brucato, J.F. Beltran, A. Abouzied, *Proceedings of the Vldb Endowment*, **9**, 576-587 (2016)
13. Y.J. He, L.H. Wu, J.C. Ren, *Introduction to Range Measurement and Control* (SDU Press, Jinan, 2009)

Research on Laser Dazzle Simulation Using ZEMAX Software and MATLAB

Niu jin[1,2,*], *Xu xiping*[1], *and An zhiyong*[1]

[1]School of Optoelectronic Engineering, Changchun University of Science and Technology, China
[2]Weaponry Equipment Research Institute of China Ordnance Equipment Group, China

Abstract. The dazzling process of the laser has been digitally simulated.First, the optical design software (ZEMAX) is combined with the scientific programming language (MATLAB), then the ray tracing is used to build the scattering model of each optical element, and finally compared with a simpler model based on CIE disability glare data to achieve the calibration and verification of the software simulation effect.The results show that this advanced optical eye simulation technology can be used to study the laser glare efficiency, and it is possible to expand the scope of application of the analysis model.

1 INTRODUCTION

Laser dazzle gained public awareness due to the increase in availability of visible wavelength lasers with powers well above laser pointers traditionally used for presentations. It has been very popular in the civil field [1,2].

The standard International Commission on Illumination (CIE) equation is a collection of research and development results of different researchers [3, 4], and has been calibrated by McLin et al. [5] for laser glare performance and integrated by Williamson. A visual tool for blinding performance research. These models can accurately simulate the dazzling effect in different situations, but they are not suitable for studying the addition of different optical components or simulating a specific real scene containing all the contributing parameters.

2. Laser dazzling simulation modeling

The human eye is one of the most complex organs of the human body. Although optically it can be considered a simple refractive system, the way the different components behave as different objects are observed, at different distances, under different illumination, and how the elements change with aging(and from person to person) poses real challenges to researchers.

2.1 Ray tracing modeling for simulating human eye design

A typical optical model is implemented using an optical design program with continuous ray tracing. That is, trace light from one surface to another in a sequential manner. Although this method can use fast and accurate tools to design and analyze optical systems, when simulating the human eye, sequential ray tracing does not fully consider an important phenomenon: scattering. To account for scattering, a common method is to use non-sequential ray tracing. This technology is based on the Monte Carlo method, which allows automatic light splitting to allow light to contact the surface in any order and number of times. Using non-sequential ray tracing, you can also define light sources and detectors in a more practical way, and users are not limited to the sequential placement of elements. However, both methods have advantages and disadvantages, and the specific choice depends on the overall analysis and research [6].

Generally, eye models implemented in optical design programs are "static"; that is, they do not allow dynamic interaction with users. Therefore, the most commonly used optical design software package (ZEMAX) and the popular scientific programming tool (MATLAB) are used after information exchange. Its main input items include the age of the human eye, the pupil size and the distance to the observed object, and also consider the scattering parameters of the main contributors (corneal, lens and fundus). Using this data, a set of parameters that define the eye can be obtained, and the ZEMAX eye design is created.

The human eye simulation module has two interfaces for entering/changing relevant parameters during the design process. As shown in Fig. 1, the human eye simulation module is responsible for defining the optical design of the human eye based on the designer's input. In its start interface, the subject's age, pupil size and distance from the object to be observed are defined. The second interface (parameterized interface) allows designers to change these parameters when necessary (for example, to add gradient index properties to the lens, consider eccentric concave positions, and even apply parameters of other eye models that they think are more

* Corresponding author: niujin0615@163.com

appropriate). From this interface, the designer can also start the optical design of the eye by entering the automatically created ZEMAX file or advancing to a specific module that can be developed using the same principle to analyze different phenomena related to optical simulation.

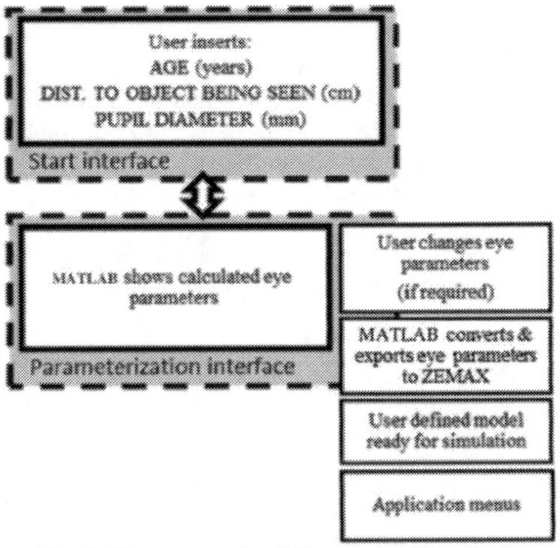

Fig. 1. Human eye simulation model flow chart

2.2 Laser dazzling scene simulation modeling

From the content of 2.1, you can get a set of parameters that define the eyes, and created the ZEMAX eye design. In order to simulate the real laser glare effect, an image model for superimposing the laser pattern on the digital image was also established.

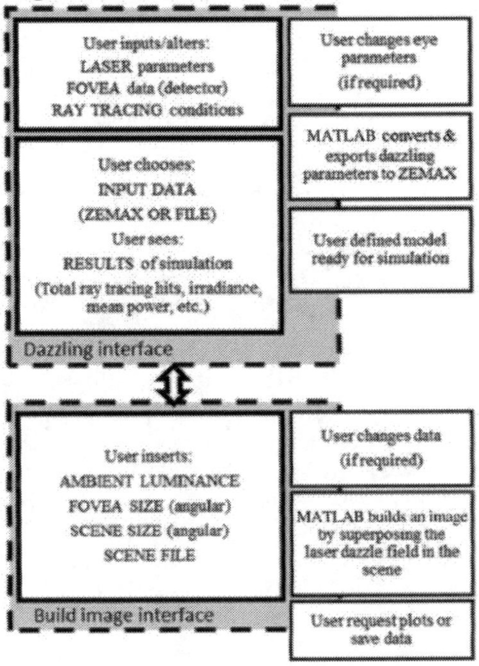

Fig. 2. Laser Dazzling Image Model flow chart

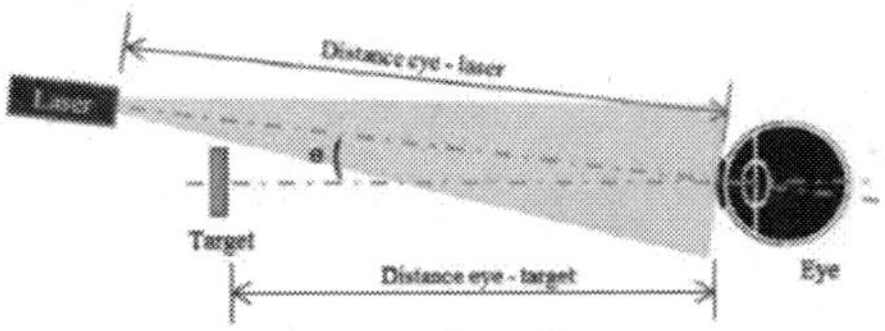

Fig.3. Schematic illustrating a laser dazzle scenario

Fig.2 illustrates the process of the laser dazzling image model, **Fig.3** shows a simplified schematic diagram of the laser dazzling scheme, and **Table 1** lists the necessary inputs to define this scheme.In the first interface of the laser glare module (the glare interface), the user can enter/change the laser parameters, the characteristics of the detection element in the retina (imaging area) and the ray tracing conditions. Designers can also import data from previous simulations (or save the current simulation). The program then allows the user to export all parameters to ZEMAX and perform ray tracing in it. Using the data obtained after ray tracing, the MATLAB-ZEMAX toolbox was used to obtain an irradiance matrix (a table of illuminance values for the entire retina) and give a set of results.

Table 1. Input table when defining laser glare scene

Element	Parameter Symbol	Units
Laser	Wavelength	nm
	Power	w
	Divergrnce	mrad
	Beam diameter	mm
	Distance to eye	m
	No. Of analysis rays	/
	Transverse or sagital angle	deg
Scene/ Target	Background luminance	cd/m^2
	Field of view	deg
Eye	Age	years
	Accommodation(dist. To taeget)	m
	Pupil diameter	mm
Imaging area(for dazzle field)	Size	mm
	No. Of pixels	/
	Field of view	deg

With the data obtained, the next task is to build a graph to load the image and overlay the laser pattern. When we input/change the ambient brightness of the scene, the FoV (angle size) of the detector and scene, and the scene file, and use the irradiance matrix from the simulation to create a laser glare image, as shown in Figure 4.

Fig. 4. Laser dazzling image overlay rendering

Convert the image irradiance to brightness, and then create an 8-bit pixel level from 0 (black) to 255 (saturated) dazzling outline, you can add the image to the scene; the saturation of the image is determined by the instantaneous dynamic range of the eye, That is, when the ratio of laser pixel brightness to ambient brightness is equal to or greater than this dynamic range, the pixel is considered saturated; if the ratio is below the reciprocal of the eye range, the pixel does not contribute anything to the final image. By assigning a value of (ratio) × (255 / eye range) to these pixels, a smooth transition can be achieved between these two levels.

3.Establishment of human eye scattering simulation model

Contrary to the commonly used parameters of the eye (curvature, distance, etc.), the scattering data on the various components of the human eye are few and incomplete. Nevertheless, this article takes a lot of empirical values to use the human eye scattering data introduced by MATLAB routines, and will analyze the scattering data of each eye element.

Massive scattering of optical elements in ZEMAX is a phenomenon that requires the use of its discontinuous mode. Two models already provided in ZEMAX can be used: Rayleigh and Henyey - Greenstein. The premise of using these two models is that the scattering length (the reciprocal of the absorption coefficient in millimeters) and the transmittance of the medium must be specified (to simulate the loss due to absorption in each scattering event). For Rayleigh scattering, the reference wavelength should be added, and the Henyey-Greenstein model requires the anisotropic coefficient g. The latter represents the average cosine of the scattering angle and varies between -1 and 1. For $g>0$, forward scattering dominates, and for $g<0$, backscattering dominates.

3.1. Cornea

Although it can be considered that the first element of the eye (the cornea) is highly transparent, transmission loss must be considered, such as loss due to scattering effects.

The initial transmission loss is due to reflection at the cornea-air interface. Because the refractive index of the cornea is considered to be about 1.377, the reflectivity at this interface can be assumed to be 2.5%, although for light with a higher angle of incidence, the reflectivity can be increased to 5%. At the cornea-water interface, the reflectivity is 2 orders of magnitude smaller.

Several experimental measurements have been reported regarding transmission. Two measurements can be considered: direct transmission and total transmission. The latter considers the overall transmittance, combining diffuse and collimated light, while direct transmission regards only collimated light (small angle transmission). In 1994, van den Berg and Tan [7] collected a set of experimental data from several authors and created a model for corneal spectral transmittance:

$$T(\lambda) = 10^{-0.016-c\lambda^{-4}} \quad (1)$$

where c represents a constant depending on the type of measurement being considered, and the wavelength, λ, is in nm. If the acceptance (full) angle of the detecting element is ~1 deg (direct transmission), then

$$c = 85\times10^{8}\,nm^{4} \quad (2)$$

while if that angle is ~180 deg (total transmission), then

$$c = 21\times10^{8}\,nm^{4} \quad (3)$$

Figure 5 shows the resulting transmittance values. Also important are the observations that, in general, the shape of the spectral transmission seems independent of age .

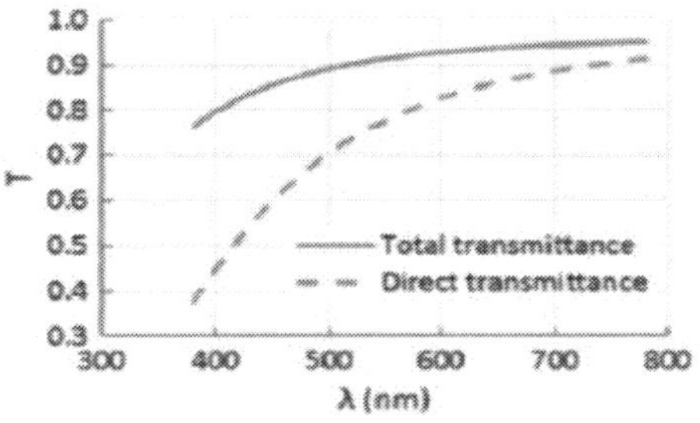

Fig. 5. Corneal spectral transmittance as calculated using Eq. (1)

Besides knowing the degree of transparency of the cornea, the simulation of light propagation through this element requires knowledge of the scattering coefficient, αs (considering that scattering is the main mechanism involved). The relationship between transmittance and scattering (e.g., [8]) is given by

$$T_s(z) = \exp(-\alpha_s z) \quad (4)$$

where α_s is the product of the cross section of the scatterers, σ_s,and their density, ρs, that is, $\alpha_s = \sigma_s \rho_s$.

If several scatterers are present, the overall scattering coefficient is obtained by summing the different scattering coefficients.

Although the cornea's composition is complex, most models for light transmission through the human cornea consider a monodisperse system of fibrils (long dielectric nonabsorbing cylinders) [9,10]. Under this assumption, Doutch et al. developed a model based on knowledge of fibrils characteristics measured by OCT, x-ray diffraction, and electron micrographs. Based on their formulas, and considering just the cornea's central position, we obtained the spectral variation of the expected scattering coefficient (Fig. 6).

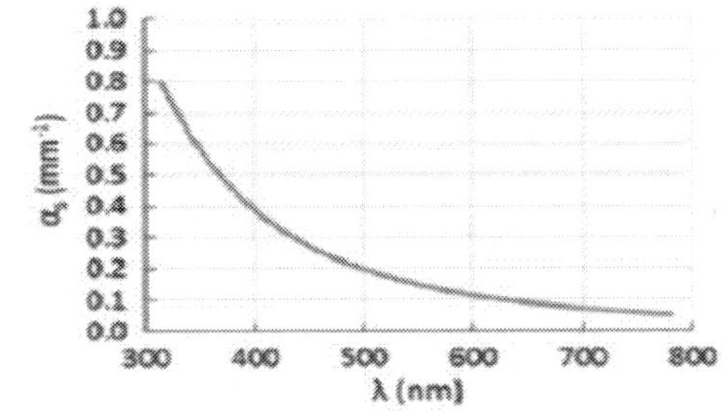

Fig. 6. Spectral variation of the cornea's scattering coefficients and length, based on the model developed by Doutch et al. [10].

In our simulator, the necessary parameters are the scattering length and the transmittance. The latter is used to simulate losses due to absorption in each scattering event. To test this implementation, we simulated only the cornea (and the aqueous humor) irradiated by a monochromatic (single wavelength), collimated, source of light with 1 W power and 2 mm beam diameter. We traced 5 × 107 rays (having a higher number did not change the data) and measured the power at the exit of the cornea with a 750 × 750 pixel detector.

Fig. 7 compares the transmittance obtained from the different models, including the implementation in ZEMAX (dEYEnamic simulator). The difference between the data from Doutch et al. [29] and that of our implementation is about 3%, which is acceptable considering that the cornea's thicknesses are different (ours is 24 μm thicker); in reality, this value is not constant as it increases from the center to the periphery (a phenomenon we did not consider). Nevertheless, our results are in accordance with the experimental average transmission presented by this model (about 2.5% less transmittance at 500 nm) and within the error stated for their data (2%) .

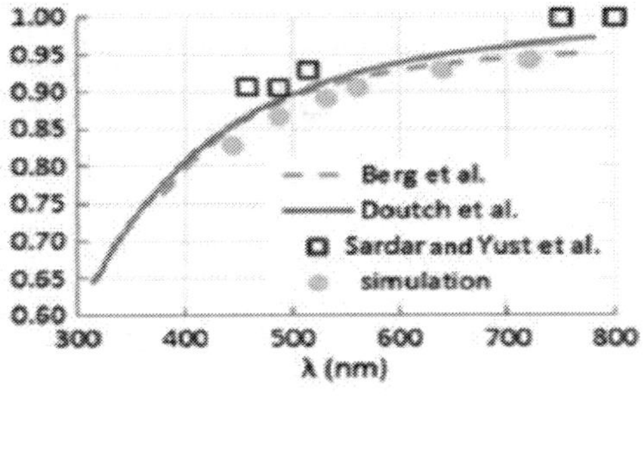

Fig.7. Simulation data of human cornea total transmittance[11-16]

3.2 lens

In addition to being dependent on the wavelength of light, the optical parameters of the human eye lens also depend on age and other factors. However, from these studies, it is possible to determine the spectral transmittance values of lenses of different ages based on the entire data set.

We derived a function to describe the increase of optical density $(\log D = 1/T_a)$ with age, based on Moreland's data. This function allows the optical density of an average observer between the ages of 20 and 60 years to be estimated using

$$D_L(\lambda, age) = D_{L1}(\lambda) \times [1.00 + 0.02 \times (age - 32)] + D_{L2}(\lambda) \quad (5)$$

while over 60 years, the equation is

$$D_L((\lambda, age) = D_{L1}(\lambda)(\times [1.56 + 0.0667 \times (age - 60)] + D_{L2}(\lambda) \quad (6)$$

where D_{L1} and D_{L2} are two optical density contributions, not varying with age (**Fig.7**). The latter only contributes for wavelengths lower or equal to 450 nm.

The more recent study of Artigas et al.[16] reported the transmittance of in vitro lenses obtained from eyes used for corneal transplants, without presenting any pathology. From their observations and in accordance with several other studies, the variation in transmission with age is only significant for elderly eyes. Above 60 years old, transmission decreases according to

$$T'(\text{age}) = -2.3452 \times 10^{-2} age + 2.3445 \quad (7)$$

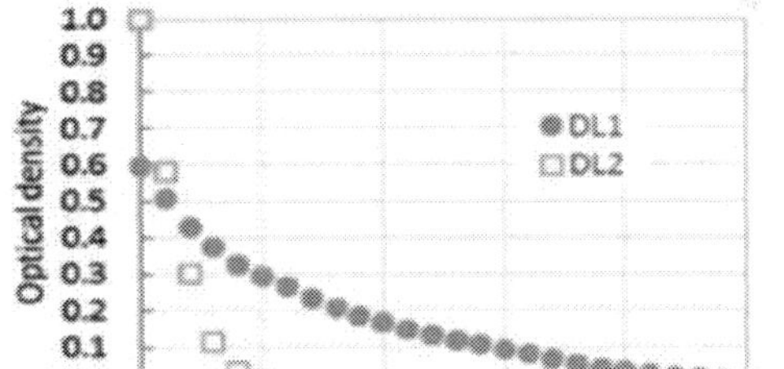

Fig.8. Optical density data contributing for Eqs.(5) and (6)

Depending on the age range being considered, we include the spectral variation by fitting data to define the following equations:

For age≥60,

$$T(\lambda, age) = 0.5[1 + erf(\frac{\lambda - 405}{63\sqrt{2}} - 1.05)]T'(age) \quad (8)$$

For 60<age≤40,

$$T(\lambda, age) = T(\lambda) = 0.5[1 + erf(\frac{\lambda - 470}{45\sqrt{2}})] \quad (9)$$

And,for age<40,

$$T(\lambda, age) = T(\lambda) = 0.5[1 + erf(\frac{\lambda - \mu}{\sigma\sqrt{2}})] \quad (10)$$

where μ =370 nm and σ=50 nm if λ > 400 nm, otherwise μ= 402 nm and σ =10 nm. The wavelength in these formulas is in nm. As before, the calculated results should be considered only as an estimate. To illustrate, according to [16], the standard deviation for adult subjects (younger than 60 years old) resides between 6% and 16% and for elderly subjects (older than 60 years old) between 6% and 21%.

4. CONCLUSIONS

The increasing number of laser dazzle events has motivated the study of glare impacts on the eye. CIE models for disability glare already quite accurately represent dazzle effects on the eye.

However, these models are based on analytical equations and have some limitations when representing the impact of the different parameters of the eye on the phenomena. Also, if one wants to develop and/or analyze the performance under laser dazzle of optical systems, they are not suitable for the task.

In order to contribute to the development of a different kind of laser dazzle simulator, we developed an optical simulator of the human eye (dEYEnamic simulator), combining the use of optical design software (ZEMAX) with a scientific programing language (MATLAB) through a dedicated toolbox. The developed interface allows the user to implement and analyze aparticular dazzle scenario using effective field parameters such as the distances from the laser to the eye, the level of accommodation of the eye (by indicating the distance to a particular object being observed), laser divergence, beam size, etc. To accomplish this, an existing optical model of the eye was upgraded with scattering data found in the literature.

Contrary to usual glare analysis, our model requires knowledge of the scattering model and parameters of each specific element of the human eye. The data are somehow scarce and,as is common with biological data, presents a high dispersion of values. This leads us to present a detailed review of the state of the art on the knowledge of these parameters and, as far as possible with the existing literature, to validate its implementation in ZEMAX.

References

1. C. A. Williamson and L. N. McLin, "Nominal ocular dazzle distance (NOOD)," Appl. Opt. 54, 1564–1572 (2015).
2. LaserPointerSafety.com, "US: FAA-reported laser incidents up significantly in 1st half of 2015," July2015
3. CIE Equations for Disability Glare Vienna: Commission Internationale de I'Eclairage (CIE, 2002).
4. J. J. Vos, "On the cause of disability glare and its dependence on glare angle, age and ocular pigmentation," Clin. Exp. Optom. 86, 363–370 (2003).
5. L. N. McLin, P. A. Smith, L. E. Barnes, J. R. Dykes, T. Kuyk, B. J. Novar, P. V. Garcia, and C. A. Williamson, "Scaling laser disability glare functions with 'k' factors to predict dazzle," in International Laser Safety Conference, Orlando, USA, 21 March 2013 (Laser Institute of America, 2013).
6. M. Stevenson, "Optical software: which program is right for me?" Opt. Laser Eur. 141, 29–32 (2006).
7. T. van den Berg and K. Tan, "Light transmittance of the human cornea from 320 to 700 nm for different ages," Vis. Res. 34(11), 1453–1456 (1994).
8. M.H.Niemz,Laser-Tissue Interactions:Fundamentals and Applications (Springer, 2007).
9. V. Tuchin, "Light scattering study of tissues," Phys.–Uspekhi 40, 495–515 (1997).
10. J. Doutch, A. Quantock, V. Smith, and K. Meek, "Light transmission in the human cornea as a function of position across the ocular surface: theoretical and experimental aspects," Biophys. J. 95, 5092–5099 (2008).
11. T. van den Berg and K. Tan, "Light transmittance of the human cornea from 320 to 700 nm for different ages," Vis. Res. 34(11), 1453–1456 (1994).
12. M. H. Niemz, Laser—Tissue Interactions: Fundamentals and Applications (Springer, 2007).
13. D. Sardar, B. Yust, F. Barrera, L. Mimun, and A. Tsin, "Optical absorption and scattering of bovine cornea, lens and retina in the visible region," Lasers Med. Sci. 24, 839–847 (2009).
14. D. Sardar, R. Yow, G. Swanland, R. Thomas, and A. Tsin, Optical Properties of Ocular Tissues in the Near Infrared Region, HumanEffectiveness Directorate, Directed Energy Bioeffects Division (Air Force Research Laboratory, 2005).
15. B.Yust,L.Mimun,and D. Sardar, "Optical absorption and scattering of bovine cornea, lens, and retina in the near-infrared region," Lasers Med. Sci. 27, 413–422 (2012).
16. J. Artigas, A. Filipe, A. Navea, A. Fandião, and C. Artigas, "Spectral transmission of the human crystalline lens in adult and elderly persons: color and total transmission of visible light," Invest. Ophtalmol. Visual Sci. 53, 4076–4084 (2012).

Study on polishing slurry of hydrogen peroxide-oxalic acid in CMP 304 stainless steel

Jianchun Weng[1, *], *Rongchuan* Lin[1], and *Xing* Rong[1]

[1]College of Mechanical and Energy Engineering, Jimei University, Xiamen 361021, China

Abstract. Stainless steel will become the substrate material of the flexible display, requirements of the flexible substrate in the surface quality and performance are very strict. Chemical mechanical polishing (CMP) is one of the most appropriate technologies to achieve the surface processing of ultra-thin stainless-steel flexible display substrate with ultra-smooth and damage-free. In this paper, the design of CMP slurry of 304 stainless steel on the hydrogen peroxide-oxalic acid type was proposed. Through experiment and analysis, the basic ingredients of CMP slurry was obtained. The research results showed that the hydrogen peroxide can increase the hydrophilicity of the stainless steel surface, and the Fenton type Haber-Weiss reaction can occur on the surface of the fresh metal substrate of stainless steel. The trivalent iron ions generated by the Fenton type reaction combined with oxygen to generate iron oxide and promoted the removal of the surface material. Under acidic conditions, the oxalic acid can decrease the stability of the oxide film on the stainless steel surface, promote the diffusion of oxygen into the metal interface, improve the oxygen reduction reaction, and increase the material removal rate. The results will provide an important reference for the next researching the CMP slurry of 304 stainless steel.

1 INTRODUCTION

Flexible display with ultra-thin, light weight, durable, large storage capacity, design freedom, flexible, winding and impact resistance and other properties [1-2], will become an important goal for the future development of display technology and widely used in industrial, civilian and military industries [3-4]. Flexible display huge market prospects, so that many countries and regions of research institutions and manufacturers into the research and application of flexible display technology [5-6], in recent years, many enterprises have introduced a foldable or bendable OLED screen [7].

Flexible display is a flexible material as substrate, the requirements on the surface quality and properties of the flexible substrate are very strict, the surface roughness must be less than 5 nm, waviness is less than 0.1 microns, high thermal stability, light weight, high strength thin, high flexibility and toughness, etc., therefore, the stainless steel material cost is low, will become the main future flexible large size display substrate materials[8-9]. The biggest problem of stainless steel as the substrate of flexible display is that the surface roughness is too large, which requires ultra-precision polishing of its surface. The processing quality and accuracy of ultra-thin stainless steel sheet will directly affect the performance of its devices[10]. At present, there is no technology for large-area (roll-to-roll) ultra-precision polishing of ultra-thin stainless-steel sheets. Therefore, how to efficiently obtain high quality and high precision in processing large-size flexible display substrates to meet the requirements of current and future flexible displays is an urgent matter for the flexible display industry[11].

Chemical mechanical polishing (CMP) technology is considered to be the best process method that can meet both surface roughness and surface smoothness requirements. It has become one of the most practical technologies for hard and brittle crystal materials to achieve the super-smooth and non-damage processing surface, and has been widely used in ultra-large-scale integrated circuits, semiconductor lighting and other fields [12]. Chemical mechanical polishing technology is probably the most suitable and fully applicable for high-efficiency ultra-precision machining of large ultra-thin stainless-steel flexible display substrate surfaces to obtain ultra-smooth and non-damaged machining surfaces.

Polishing slurry is an important component in chemical mechanical polishing. Its cost accounts for 60%~70% of the total cost of chemical mechanical polishing, so its quality determines the efficiency, surface quality and cost of chemical mechanical polishing. Oxidant and pH regulator are the important components in a chemical mechanical polishing slurry, which can promote the chemical reaction of the polished surface material, make the abrasives play the micro cutting function evenly and effectively, and improve the material removal. Therefore, it is urgent to study the chemical mechanical polishing slurry of stainless steel with environmental protection and high efficiency.

Therefore, in order to develop an environmentally friendly and efficient chemical mechanical polishing

* Corresponding author: gingerwjc@163.com

slurry for stainless steel, this paper proposes the design of hydrogen peroxide-oxalic acid type 304 stainless steel chemical mechanical polishing slurry, which provides theoretical and technical support for the researches of large-scale production, chemical mechanical polishing slurry and chemical mechanical polishing mechanism.

2 Experimental preparation and experimental parameters

All the chemical mechanical polishing experiments conducted in class 1000 clean room, the environment temperature control in 22°C, the resistivity of demonized water used in the experiment is 18.24 MΩ·cm. Samples used in chemical mechanical polishing experiment were a number of 50 mm diameters 304 stainless steel sheets. Before the experiment, its surface roughness Ra was 40-50 nm. The polishing experiment was carried out on a ZYP300 lapping and polishing machine made in Shenyang, and the type of polishing pad was Rode IC1000.

Polishing parameters selection. The rotational speed of polishing platen is set 60 r/min, the carrier speed is set for 60 r/min, the polishing pressure P is 2 psi, and the polishing time is set 15 min. Every time after polishing, conditioning the polishing pad and dressing time is for 15 min. In the process of polishing, the carrier along the arc reciprocating swing, swing range of 20 mm, oscillation frequency of 10 s, center distance of the object stage and pads is set to 80 mm.

Test instrument used in the experiment. Sartorius CP225D precision balance (0.01mg) was used to measure the sample weight before and after the polishing experiment, and then the material removal rate was calculated by calculation. 3D surface microscope (0.01nm vertical resolution) of Contour GT-K with BRUKER companies in the United States production used to measure the surface topography and the surface roughness of samples before and after polishing. The Lecia metallurgical microscope with DM2500M is used to measure the 2D original image of the surface before and after polishing. Using JNGX JL-1197 laser particle instrument tests the distribution of abrasive polishing and with pH (0.1) precision electronic testing pen test polishing slurry pH value.

Basic ingredients of polishing slurry. According to the results of the previous polishing slurry test and the material characteristics of 304 stainless steel, white corundum was selected as the abrasive for the polishing experiment, sodium native phosphate was selected as the dispersant, oxalic acid was used as the pH regulator, and hydrogen peroxide was used as the oxidant. The total volume of the polishing slurry was 250ml for the each chemical mechanical polishing experiment.

3 The experiment

3.1 Orthogonal experimental designs

According to the previous studies, the orthogonal test table without interaction was adopted to carry out the orthogonal experiment, which was repeated for three times. The experimental results were analyzed the significance of each component on the removal rate and surface roughness of 304 stainless steel materials through SAS software, and the optimal polishing slurry component was obtained comprehensively according to the single factor experiment results.

The factors to be determined in this orthogonal experiment include: particle size and content of white corundum, sodium hexametaphosphate, oxalic acid and hydrogen peroxide content. Four levels are selected for each factor, and the factor levels of the orthogonal experiment are shown in Table 1.

Table 1. Table of orthogonal test factors for hydrogen peroxide oxalic acid type polishing slurry.

Factor / The level of	a	b	c	d	e
	Particle size(μm)	**Particle content (g)**	**Sodium hexameta phosphate (g)**	**Oxalicacid (g)**	**Hydrogen peroxide (g)**
1	0.5	1.5	1	1	5
2	3.5	4.5	3	2	10
3	7	7.5	5	3	15
4	14	10.5	7	4	20

3.2 Orthogonal experimental table and experimental results

According to the number of factors, the horizontal orthogonal table L_{16} (4^5) with 5 factors and 4 factors ware selected. After the test, the material removal rate and surface roughness test data were recorded, and the experimental results were shown in Table 2 of the orthogonal test.

Table 2. Optimization test data of hydrogen peroxide oxalic acid type polishing slurry.

Factor / Number	a Particle size (μm)	b Amount of abrasive (g)	c Sodium hexameta phosphate (g)	d Oxalic acid (g)	e Hydrogen peroxide (g)	MRR (nm/min)	Roughness (μm)
1	0.5	1.5	1	1	5	60.19632	0.011
2	0.5	4.5	3	2	10	109.89687	0.012
3	0.5	7.5	5	3	15	136.75368	0.017
4	0.5	10.5	7	4	20	121.31873	0.017
5	3.5	1.5	3	3	20	129.65361	0.017
6	3.5	4.5	1	4	15	145.39726	0.018
7	3.5	7.5	7	1	10	149.71904	0.05
8	3.5	10.5	5	2	5	184.60204	0.03
9	7	1.5	5	4	10	145.08856	0.034
10	7	4.5	7	3	2	156.81912	0.05
11	7	7.5	1	2	20	140.14937	0.04
12	7	10.5	3	1	15	141.69287	0.023

13	14	1.5	7	2	15	148.79295	0.055
14	14	4.5	5	1	20	133.97539	0.036
15	14	7.5	3	4	5	162.37571	0.042
16	14	10.5	7	3	10	197.87610	0.101

3.3 Experimental results and analysis

3.3.1 The influence of various factors on material removal rate and surface roughness

According to the analysis of variance, factors a, b, c, d and e have a very significant impact on the material removal rate and surface roughness (because the Pr>F value of each factor is less than 0.01). The order of significance of each factor on material removal rate was: F (a) >F (b) >F (d) >F (e) >F (c); The influence degree of each factor on the material removal rate is the particle size, particle content, oxalic acid content, hydrogen peroxide content, sodium hevatophosphate content in descending order. The order of significance of each factor on surface roughness: F (a) >F (e) >F (c) >F (d) >F (b); The degree of influence of various factors on surface roughness is in descending order from the largest to the smallest, namely, particle size, hydrogen peroxide content, sodium hevatophosphate content, oxalic acid content and particle content.

3.3.2 The optimal combination of polishing slurry and polishing scheme

The optimized composition of the polishing slurry is shown in Table 3, and Table 4 is the verification test of the optimized polishing slurry. Figure 1 shows 2D and 3D diagrams of the surface topography of 304 stainless steel polished by optimized polishing slurry.

Table 3. Optimal polishing slurry obtained by orthogonal test.

Factor	a Particle size (μm)	b Grinding particle content (g)	c Sodium hexameta phosphate (g)	d Oxalic acid (g)	e Hydrogen peroxide (g)
Numerical	3.5	7.5	5	4	15

Table 4. Optimized mechanical polishing test results of polishing liquefaction.

Factor / Number	MRR(nm/min)	Ra
1	201.8794	0.018
2	205.9571	0.017
3	200.6874	0.015

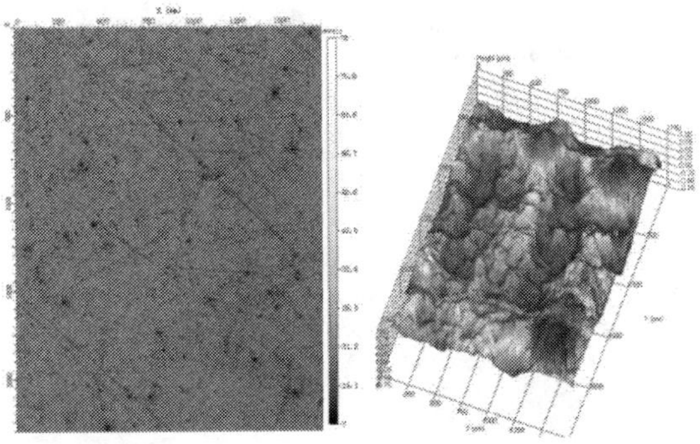

Figure 1. 2D and 3D diagrams of surface topography after polishing 304 stainless steel with optimized polishing slurry.

Figure 1 shows the surface morphology of 304 stainless steel after polishing with hydrogen peroxide oxalic acid polishing slurry. Based on the optimal ratio of polishing slurry, only the particle size was changed, and the polishing experiment was carried out under the same conditions to measure the material removal rate and surface roughness, and the polishing scheme of 304 stainless steel was obtained. See Table 5.

Table 5. Particle size selection scheme of hydrogen peroxide oxalic acid polishing slurry.

Particle diameter	**30 (nm) Silica sol**	**50 (nm) Silica sol**	**W0.5 White fused alumina**	**W3.5 White fused alumina**	**W5 White fused alumina**	**W7 White fused alumina**	**W14 White fused alumina**
MRR (nm/min)	71.86	90.60	126.75	201.88	212.37	145.39	171.57
Ra(μm)	0.007	0.009	0.017	0.018	0.023	0.038	0.040
Polishing stage	Semi finish polishing	Semi finish polishing	Rough polishing	Rough polishing	Rough polishing	Rough polishing	Rough polishing

4 Discuss

Oxalic acid ACTS as a pH regulator and hydrogen peroxide ACTS as an oxidant in the H_2O_2_oxalic acid chemical mechanical polishing slurry. When the oxalic acid content was 1.6%wt and hydrogen peroxide content was 6%wt, the material removal rate reached the maximum value. In chemical mechanical polishing, it is concluded that the addition of hydrogen peroxide can effectively improve the material removal rate.

(1) In chemical mechanical polishing, polishing interface can produce high temperature because of friction, this will increase the hydrophilicity on the surface of the stainless steel, under the action of hydrogen peroxide, this will increase the content of oxygen groups on the stainless steel surface. On the one hand, the hydrogen peroxide has the strong oxidizing, would that be to form a layer of dense oxide film on stainless steel surface [13], and it will be removed by the mechanical action of the particle. On the other hand, due to the presence of

oxygen in the polishing slurry, stainless steel will produce dissolved oxygen electrochemical corrosion[14], and the reaction formula is $Fe+1/2O_2+H2O \rightarrow Fe(OH)_2$.

(2) Oxalic acid is added into the polishing slurry to make the polishing slurry acidic. Under acidic conditions, due to the presence of H^+, on the one hand, the stability of the oxide film on the surface of stainless steel is reduced, which promotes the diffusion of oxygen to the metal interface and improves the oxygen reduction reaction; on the other hand, the hydrogen reduction reaction is increased, this is $2H_3O^++2e \rightarrow H_2\uparrow+2H_2O$, making the metal more soluble[15].

(3) Hydrogen peroxide, the oxidant in the polishing slurry, will have a Fenton-type Haber-Weiss reaction on the surface of the fresh metal matrix of stainless steel[16]. The decomposition of hydrogen peroxide will be accelerated, and the ferric ion generated by the reaction will combine with oxygen to form iron oxide. The main reaction mechanism is roughly as following:

$$Fe^{2+}+H_2O_2 \rightarrow Fe^{3+}+HO\cdot+OH^+ \quad (1)$$

$$Fe^{2+}+HO\cdot \rightarrow Fe^{3+}+HO^- \quad (2)$$

$$H_2O_2+HO\cdot \rightarrow HO_2\cdot+H_2O \quad (3)$$

$$Fe^{2+}+HO_2\cdot \rightarrow Fe^{3+}+HO_2^- \quad (4)$$

$$Fe^{3+}+HO_2\cdot \rightarrow Fe^{2+}+H^+O_2 \quad (5)$$

$$Fe^{3+}+HO_2^- \rightarrow Fe^{2+}+HO_2\cdot \quad (6)$$

The iron oxide generated by the above reaction does not adhere firmly to the surface of the metal matrix, which is easy to fall off in the friction process and form particles. The metal matrix is exposed again, and the above process occurs repeatedly. The results show that the lower the pH of hydrogen peroxide polishing slurry, the more stable the oxidant hydrogen peroxide, the higher the material removal rate and the lower the surface roughness can be obtained.

5 Conclusion

To sum up, through a series of experiments, result analysis and discussion, the following conclusions are drawn:

(1) As a polishing slurry oxidant, the hydrogen peroxide can increase the hydrophilic of the stainless steel surface, increase the content of the stainless steel surface oxygen-containing groups, oxidize the surface of the stainless steel and formation of a layer of dense oxide film. And the existence of oxygen in the polishing slurry, will also make stainless steel produced dissolved oxygen electrochemical corrosion, promote the removal of materials.

(2) Hydrogen peroxide will produce Fenton-type Haber-Weiss reaction on the surface of the fresh metal matrix of stainless steel, and meanwhile, the ferric ion generated by the reaction will combine with oxygen to form iron oxide, promoting the material removal.

(3) When the pH value regulator of hydrogen peroxide based polishing slurry is oxalic acid, environmental protection requirements can be realized while adjusting the pH value. The lower the pH value of the polishing slurry, the greater the material removal rate.

(4) In acidic conditions, it will make the stainless steel surface oxidation film stability decline, promote oxygen to the metal interface diffusion and improve oxygen reduction reaction. Due to the presence of H^+, the hydrogen reduction reaction will increase, making the metal more soluble and promoting the material removal.

(5) Through orthogonal design and variance analysis The basic formula of H_2O_2-oxalic acid chemical mechanical polishing slurry for stainless steel was obtained. According to surface quality requirements of polishing, the selection of rough polishing and fine polishing process could be realized by changing the particle size.

The results of this paper provide an important reference for the further study of chemical mechanical polishing slurry of 304 stainless steel.

Acknowledgments

Natural Science Foundation of Fujian Province, China (2019J01709). Xiamen major science and technology projects (3502Z20191022).

References

1. Logothetidis S, Laskarakis A. Towards the optimization of materials and processes for flexible organic electronics devices[J]. The European Physical Journal Applied Physics, 2009,46(1): 12502(1-9).
2. Hwang Han Wook, Hong Seonghwan, Hwang Sang Soo, et al. Analysis of Recoverable Residual Image Characteristics of Flexible Organic Light-Emitting Diode Displays Using Polyimide Substrates[J]. IEEE Electron Device Letters, 2019,40(7):1108-1111.
3. Lee Seunghwan, Han Ju-Hwan, Lee Seong-Hyeon, et al. Review of Organic/ Inorganic Thin Film Encapsulation by Atomic Layer Deposition for a Flexible OLED Display, Journal of metals, 2019,71(1):197-211.
4. Bae S, Kim H, Lee Y, et al. Roll-to-roll production of 30-inch graphene films for transparent electrodes [J]. Nature Nanotechnology, 2010, 5(8):574-578
5. Hong Kihyon, Yu Hak Ki, Lee IIhwan, et al. Flexible top-emitting organic light emitting diodes with a functional dielectric reflector on a metal foil substrate, RSC Advances,2018,8(46): 26156-26160.
6. Someya T. Flexible electronics: Tiny lamps to illuminate the body [J]. Nature Materials, 2010, 9(11):879-880.
7. Xu zheng, song dandan, zhao su ling et al. Development and challenges of OLED, flexible, transparent display technology and organic light-

emitting materials. Modern display, 2009,101 (6) : 5-10

8. Sugimoto Akira, Ochi Hideo, Fujimure Soh, et al. Flexible OLED displays using plastic substrates [J]. IEEE J Sel Top Quantum Electron, 2004, 10(1): 107-114.
9. Bardsley J N. International OLED technology roadmap [J]. IEEE J Sel Top Quantum Electron, 2004, 10(1): 3-9.
10. Xie zhiyuan, Hung Liangsun, Zhu Furong. A flexible top-emitting organic light-emitting diode on steel foil[J]. Chem Phys Lett ,2003,381(5/6):691-696.
11. Liu haiyan. Performance and testing of glass substrates for tef-lcd [J]. Glass, 2009,36(1):22-24
12. Xiaokai Hu,Zhitang Song,Fei Qin,et al. Chemical mechanical polishing of stainless steel foil as flexible substrate. Applied Surface Science, 2012 ,258(15):5798-5802.
13. XIAO Xin, ZHONG Ping. Rapid pickling for stainless steel at room temperature, Electroplating & Finishing, 2005,24(1) 65-66.
14. Bao Qinai.The relationship between chloride ion and the corrosion of stainless steel in cooling water system,Industrial Water Treatment,2007,27(9) 1-6.
15. YE Chao, DU Nan, TIAN Wenming, et al. Effect of pH on Pitting Corrosion Process of 304 Stainless Steel in 3.5%NaCl Solution, Journal of Chinese Society for Corrosion and Protection, 2015, 35(1)38-43.
16. Wang Jihui, Gu Kali, Wuli. Surface Passivation Treatment and Tribological Properties of 1Cr18Ni9Ti Stainless Steel in Hydrogen Peroxide Solution, Tribology, 2009,29(1):18-26

Simulation of Electrical Characteristics on Inhomogeneous Strains in Normally-off HEMTs with p-GaN Gate

Jing Zhou[1,*], *Hongjun* Chen[1], *Yang* Wang[1], and *Xingming* Long[2]

[1]Full State Key Lab of Power Transmission Equipment and System Security & New Technology, Chongqing University, Chongqing 400047 China
[2]Department of Physics, Chongqing Normal University, Chongqing 401331 China

Abstract. Strain is one of the important factors affecting the two-dimensional electron gas (2DEG) transform in AlGaN/GaN material based high electron mobility transistors (HEMTs) by polarization effects. In this paper, the effects of inhomogeneous biaxial strain in different regions of the AlGaN barrier layer on electrical properties of normally-off HEMTs with p-GaN gate were discussed. The results show that biaxial strain applied in three regions has different influence on transfer, output and breakdown characteristics of the device. The strain applied in region under gate has the most significant impact on threshold voltage and drain saturation current with a decreasing of 39% and an increasing of 97% respectively as the strained lattice constant increases from 3.173061Å to 3.187229Å.While, strain applied between gate and drain electrode can improve the off-state breakdown voltage by 12% with the increasing of strained lattice constant.

1 Introduction

High Electron Mobility Transistors (HEMTs) based on AlGaN/GaN heterostructures became attractive candidates for high switching frequency, high breakdown voltage and high power due to their excellent electrical properties, compared to their silicon (Si) counterparts [1-3]. In the interface of heterojunction, a two dimensional electron gas (2DEG) with a high sheet carrier density ($\sim 10^{13}cm^{-2}$) and mobility (above $1000cm^2V^{-1}s^{-1}$) is generated, owing to the spontaneous and piezoelectric polarization, and donor-like AlGaN surface states acting as source of electrons. Generally, most AlGaN/GaN HEMTs exhibit normally-on operation since 2DEG is intrinsically existed in the AlGaN/GaN interface when the HEMT is operating at a gate bias of less than zero volts [4-5]. However, for power electronics applications normally-off devices is needed considering of safety conditions and simplicity in the gate driver circuits.

Several approaches have been reported to obtain normally-off GaN-based HEMTs, covering the Recessed gate, Fluorine ions injection, and Recessed Hybrid MIS (metal-insulator-semiconductor) HEMT configurations. However threshold voltage (V_{th}) stability can be a safety concern in all this kind of devices [6-7]. To date the most typical way to yield a normally-off device is the use of a p-GaN gate, grown on the AlGaN/GaN heterostructure only in the gate contact region, which can deplete the conduction channel when unbiased.

Previous studies demonstrate that 2DEG can be affected by strain [8]. Tong et al. [9] applied a strain model to study the electron transport in normally-on HEMT device when barrier is uniformly biaxial strained. Simin G. et al [10] revealed that gate bias-induced inhomogeneous strain in the AlGaN barrier will cause a decrease in polarization charge and reduction in 2DEG in normally-on HEMT device by experiments. Ahmeda K. et al [11] found that applying strain in drain region will cause changes in the total polarization, thus affecting 2DEG density in the channel. However, as far as we known, there are few reports on the strain effect in the AlGaN barrier of normally-off HEMT with a p-GaN gate. And the effect of non-uniform strain on device characteristics is often omitted because the strain is treated as uniform directly.

In this work, numerical simulations of the effects of inhomogeneous biaxial strain in AlGaN barrier layer on the electrical properties in a normally-off HEMTs with p-GaN gate are discussed. The biaxial strain in three regions, region under the gate (Region-G), region between the gate and the drain electrode (Region-GD) and region between the gate and the source electrode (Region-GS) are found to change with the strain in the lattice constant. And the impact of strain on electrical parameters, such as threshold voltage (V_{th}), drain saturation current (I_{sat}) and off-state breakdown voltage (V_{br}) is simulated and comparatively investigated using Sivalco-TCAD-altas.

2 Structure Model and Analysis Method

The schematic diagram of normally-off HEMT with p-GaN gate is presented in Fig.1. The S, G, and D denotes the source electrode, the gate electrode and the drain electrode respectively. The Nitride (Si_3N_4) is the

* Corresponding author: jeanne_z@cqu.edu.cn

passivation layer, with which the surface state of AlGaN barrier can be improved to suppress the current collapse phenomenon. The $Al_{0.07}Ga_{0.93}N$ buffer layer allows a better confinement of 2DEG in the channel [12,13]. The structure parameters used in the simulation are listed in Table 1.

The relationship between strain and 2DEG is defined by a strain model [9]. When the strained lattice constant (a) in AlGaN barrier layer changes, the external biaxial strain will change correspondingly. The total polarization charge, including charge of piezoelectric polarization and spontaneous polarization, changes as well, and ultimately affecting the concentration of 2DEG. Coupling with Poisson's equation and current continuity equation, the "*I-V*" characteristics of the normally-off HEMT with p-GaN gate can be obtained.

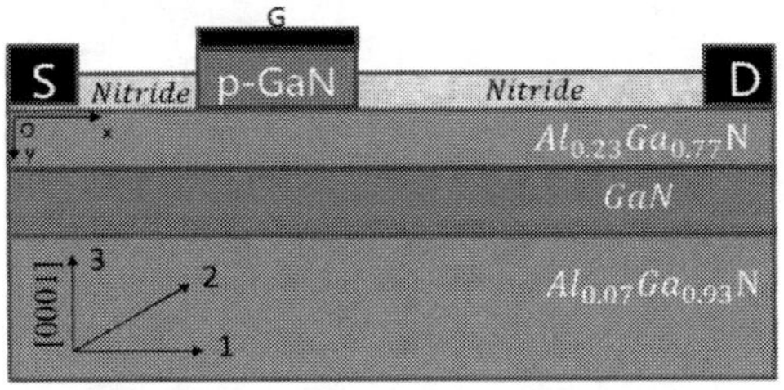

Fig. 1. Schematic diagram of the normally-off AlGaN/GaN HEMT with p-GaN gate

Due to the growth of two piezoelectric materials GaN and AlGaN layers, a lattice mismatch strain is developed in the heterostructure.as shown in Eq.(1), the strains developed in the AlGaN layer due to lattice mismatch are:

$$\varepsilon_1=\varepsilon_2=\frac{a_{GaN}-a_{AlGaN}}{a_{AlGaN}},\quad \varepsilon_3=-2\frac{C_{13}}{C_{33}}\varepsilon_1 \qquad (1)$$

Where a_{GaN} and a_{AlGaN} are the lattice constant for GaN and AlGaN without external strain, respectively and C_{13} and C_{33} are the elastic constants of AlGaN layer. Fig.2 illustrates the state of normally-off HEMT under external biaxial strain and the external strain for AlGaN can be defined as Eq.(2):

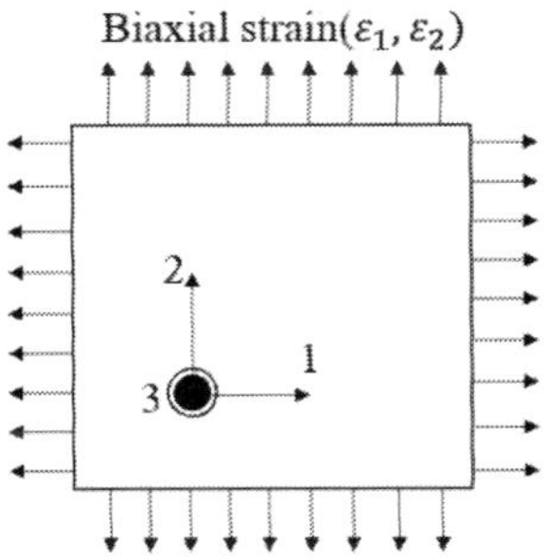

Fig. 2. Top-view of the normally-off AlGaN/GaN HEMT under biaxial applied strain in the 1- and 2-directions (basal plane)

Where a and $\boldsymbol{a_0}$ represent the strained and unstrained lattice constant of AlGaN barrier layer respectively, and $(a-a_0)/a_0$ is restricted to a small range [13] less than 0.06% according to the stain model [9] on which the investigations are based.

Table 1. Device structure parameters used in simulation

Device parameters	Value
Length of source, L_s	1 μm
Length of gate, L_g	1.4 μm
Length of drain, L_d	1 μm
Source-Gate spacing, L_{gs}	1 μm
Gate-Drain spacing, L_{gd}	6 μm
p-GaN thickness	35 nm
p-GaN uniform doping	$3\times10^{17}cm^{-3}$
$Al_{0.23}Ga_{0.77}N$ barrier thickness	15 nm
GaN thickness	10 nm
$Al_{0.07}Ga_{0.93}N$ buffer thickness	2 μm
Si_3N_4 thickness	0.31 μm
Gate width	1000 μm

Assuming the GaN layer is under total strain relaxation and $\boldsymbol{P_{SP}}(\mathbf{GaN})$ is a constant unaffected by strain, the polarization induced charge density can be calculated by Eq. (3) [9].

$$\begin{aligned}\sigma&=P_{SP}(GaN)-P_{SP}(AlGaN)-P_{PE}(AlGaN)\\&=(-0.034)-(0.056x-0.034)-(\varepsilon_1+\varepsilon_1^{'})e_{31}\\&\quad-(\varepsilon_2+\varepsilon_2^{'})e_{31}-(\varepsilon_3+\varepsilon_3^{'})e_{33}\\&=(-0.034)-(0.056x-0.034)\\&\quad-2\frac{a-a_0}{a_0}(e_{31}-e_{33}\frac{C_{13}}{C_{33}})\end{aligned} \qquad (2)$$

Where $\boldsymbol{P_{SP}}$ and $\boldsymbol{P_{PE}}$ denote spontaneous polarization and piezoelectric polarization respectively. The spontaneous polarization on the GaN and AlGaN layer ($\boldsymbol{P_{SP}}(\mathbf{GaN})$) and ($\boldsymbol{P_{SP}}(\mathbf{AlGaN})$) is an intrinsic parameter[13], x is the Al alloy composition of $\boldsymbol{Al_xGa_{(1-x)}N}$ barrier, e_31 and e_33 are the piezoelectric coefficients of AlGaN layer, $\mathbf{C_{31}}$ and $\mathbf{C_{33}}$ are the elastic constants of AlGaN layer[15] Then the density of 2DEG in the channel can be given by[14] Eq. (4).

Where $\boldsymbol{\sigma}$ is the total polarization induced charge density defined by Eq. (3), $\boldsymbol{\epsilon}$ is the relative dielectric constant of AlGaN layer, $\boldsymbol{t_{AlGaN}}$ is the thickness of the barrier layer, $\boldsymbol{\varphi_b}$ is the Schottky barrier of a gate contact, $\mathbf{E_F}$ is the Fermi level regarding to the GaN conduction-band-edge energy in the interface and $\mathbf{\Delta E_c}$ is the band offset between AlGaN and GaN interface.

In this paper, the electrical characteristics under external inhomogeneous biaxial strain are studied by adjusting the strained lattice constant (a) in different regions of barrier separately. The $\boldsymbol{a_0}$ is set to 3.17129Å [15], and the sets of strained lattice constant a are

3.173061Å , **3.176603Å**, **3.180145Å**, **3.183687Å**, and **3.187229Å**, with a step length of **0.003542Å**.

3 Discussion

The electrical characteristics of the device were shown in Fig.3 when the *a* of AlGaN barrier in Region-GS increases while the *a* of the other two regions keeps the same as before (3.189Å). Fig.3(a) demonstrates the density of 2DEG increases with *a* which is in coherent with Eq.(3). Fig.3(b) shows the transfer characteristics of the normally-off HEMTs, where the gate-source voltage V_{gs} varies from -2V to 8V and the drain-source voltage V_{ds} is 7V..

The HEMT, having $Al_{0.23}\,Ga_{0.77}\,N/GaN$ with a p-GaN gate, shows a typical normally-off operation with a threshold voltage above 1V, as defined by the gate bias intercept of a linear extrapolation of drain current at the point of peak g_m (not shown in Fig.3).Strain-enhanced V_{th} can be derived from reduced density of 2DEG in the Region-GS, and the gate voltage required to open the channel shift positively, which agrees with the results from choi[16]and Liu[17]. As shown in Fig.3(b), V_{th} increases from 1.4V to 2.1V with a rate up to 37% when *a* decreases from 3.187229Å to 3.173061Å

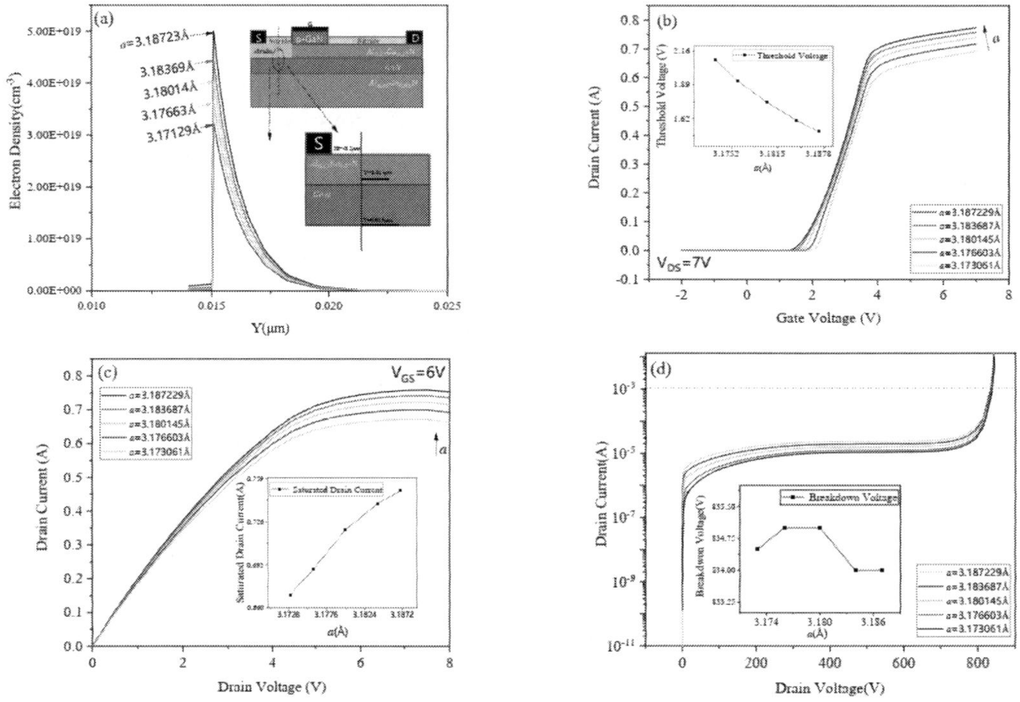

Fig. 3.(Color online) The impact of strained lattice constant *a* in Region-GS on electrical characteristics.(a) the cross-sectional 2DEG density at x= -0.5μm where the peak 2DEG density was located close to the AlGaN/GaN interface.(b) transfer characteristics.(c) output characteristics.(d)breakdown characteristics. The insets give the change of V_{th}, I_{sat}, V_{br} against strained lattice constant a.

Furthermore, the drain saturation current I_{sat}, which is defined as the maximum current in the saturated region at V_{gs}=6V and V_{ds} varies from 0 to 8V, become smaller. This reduction in I_{sat} may be caused by the lower density of 2DEG due to a smaller *a* in the Region-GS which affecting the sheet resistance and current drive, and literature [17,18] reported the similar results. I_{sat} grows about 12% when a varies from 3.173061Å to 3.187229Å. In addition, the breakdown voltage V_{br} of the device, defined as the applied drain voltage when the drain current reaches 1mA at the gate voltage of 0V with source electrode grounded, are investigated. The results in Fig.3(d) denote that local strain in Region-G has a little impact on the off-state breakdown characteristics. The effect on electrical characteristics by the strain in Region-G has a little impact on the off-state breakdown characteristics.

The effects on electrical characteristics by the strain in Region-G and Region-GD were simulated using the same method. A comparative analysis among the influence of strain working in different regions of AlGaN barrier is carried out. Fig.4(a) shows the effect of inhomogeneous strain on threshold voltages V_{th}. It is clearly observed that the threshold voltage V_{th} reduces up to 39% when a in Region-G decreases from 3.187229Å to 3.173061Å, while only 33% and 37% with the same change of a in Region-GD and Region-GS respectively. The results indicate that strain in Region-G has the most significant influence on V_{th}. In addition, the similar results had been obtained by Choi S. who found that the decreasing of total polarization charge and the 2DEG concentration resulted in the positive shifts of V_{th} by increasing the In composition($In_x\, Al_{(1-x)}\, N$) in experiments [16]. Fig.4(b) shows the influence of strain in different regions on drain saturation current I_{sat}. Strain in Region-G has the most influence on I_{sat} with an increasing of 97% when a grows from 3.173061Å to 3.187229Å. And the data collected from Liu[17] shows that external strain from compressive to tensile increases drain current, which is in line with our simulation. Fig.4(c) gives the curves of breakdown voltage against a of different regions. The breakdown voltage has an evident positive shift under the strain of Region-GD. It is claimed that increased strain in the AlGaN layer causes increasing density of 2DEG. In turn, it decreases the electric field locally[20], hence enhance the off-state breakdown voltage. And V_{br} can be improved by 12% from 731V to 821V when the strained lattice constant a changes from 3.173061Å to 3.187229Å.

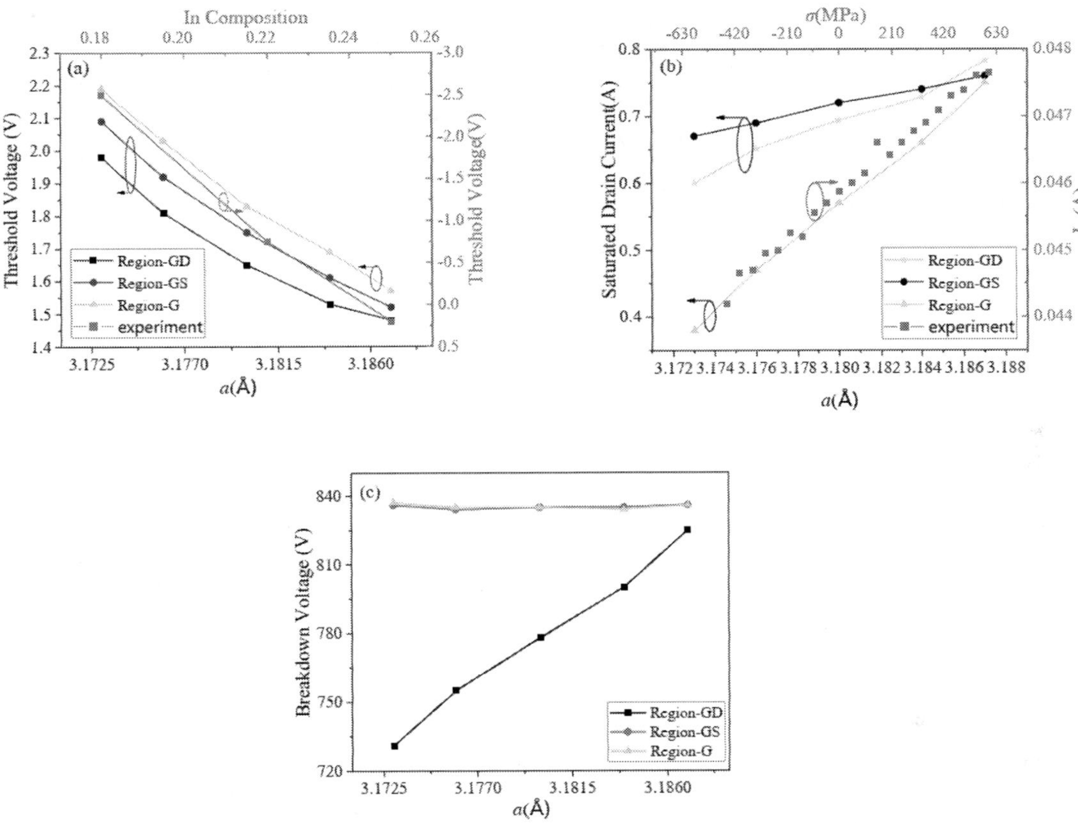

Fig. 4 (Color online) The relationship between the strained lattice constant $\boldsymbol{a}$ of Region-GS, Region-G and Region-GD in HEMT and the electrical features, compared to the experimental data[16][17]: (a) the threshold voltage $\boldsymbol{V_{th}}$, (b) the drain saturation current $\boldsymbol{I_{sat}}$, and (c) the breakdown voltage $\boldsymbol{V_{br}}$

4 Conclusion

In this paper the influence of inhomogeneous biaxial strain in barrier on electrical characteristics of normally-off HEMT with p-GaN gate have been numerically explored. The simulation results indicate that strain on the AlGaN barrier layer has a significant impact on the electrical characteristics of the HEMT device. In Region-G or gate region, the strain has the most predominant impact on V_{br} with a decreasing of 39%, while I_{sat} with an increasing of 97% as the a increases from 3.173061Å to 3.187229Å. In addition, the V_{br} can be improved with 12% by strain between the gate and drain electrode or Region-GD when the increases. The built model and results can pave the way for optimizing the performance and improving the reliability of the under developing normally-off HEMTs. should be centred and should be numbered with the number on the right-hand side.

References

1. A. Mecke, I. Lee, J.R. Baker jr., M.M. Banaszak Holl, B.G. Orr, Eur. Phys. J. E **14**, 7 (2004) [1]Greco, Giuseppe Iucolano, Ferdinando Roccaforte, Fabrizio, MAT SCI SEMICON PROC., 8, 2018.
2. W. Runton, B. Trabert, J. B. Shealy, and R. Vetury, IEEE Microw. Mag., **14**, 3,(2013)

3. J. J. Komiak, IEEE Microw. Mag., **16**, 3, (2015)
4. Roccaforte F, Fiorenza P, Greco G, Lo Nigro R, Giannazzo F, Iucolano F, et al. MICROELECTRON ENG. **187-188**. (2018)
5. Yao Y, Huang C, Lin T, Cheng L, Liu C, Wang M, et al. MICROELECTRON ENG. **138** ,(2014)
6. Saito W, Takada Y, Kuraguchi M, Tsuda K, Omura I. IEEE T ELECTRON DEV. **53**, (2006)
7. Cai Y, Zhou Y, Lau KM, Chen KJ. IEEE T ELECTRON DEV. **53**, (2006)
8. Yalamarthy AS, Senesky DG. SEMICOND SCI TECH. **31**, (2016)
9. Tong W, Tang W, Zhang Z. COMP MATER SCI. vol. **143**, (2017)
10. Simin G, Koudymov A, Tarakji A, Hu X, Yang J, Khan MA, et al. APPL PHYS LETT. **79**, (2001)
11. Ahmeda K, Ubochi B, Benbakhti B, Duffy SJ, Soltani A, Zhang WD, et al. IEEE ACCESS. **5**,(2017)
12. O. Hilt, A. Knauer, F. Brunner Proceeding of The 22nd International Symposium on Power Semiconductor Devices&ICs,(2010)
13. Benkhelifa F, Muller S, Polyakov VM, Ambacher O. IEEE ELECTR DEVICE L. **36** (2015)
14. Yu ET, Dang XZ, Yu LS, Qiao D, Asbeck PM, Lau SS, et al. APPL PHYS LETT.**73**, (1998)
15. Wright AF. J APPL PHYS. **82**,(1998)
16. Choi S, Kim HJ, Lochner Z, Zhang Y, Lee Y, Shen S, et al. APPL PHYS LETT. **96**, (2010)
17. Liu K, Zhu H, Feng S, Shi L, Zhang Y, Guo C. MICROELECTRON RELIAB. **55**, (2015)
18. Ambacher O, Foutz B, Smart J, Shealy JR, Weimann NG, Chu K, et al. J APPL PHYS. **87**, (1999)
19. Yalamarthy AS, Senesky DG. SEMICOND SCI TECH..**31**,. (2016)
20. Duffy SJ, Benbakhti B, Kalna K, Boucherta M, Zhang WD, Bourzgui NE, et al. IEEE ACCESS.**6**, (2018).

A new learning algorithm based on strengthening boundary samples for convolutional neural networks

Dongning Zhou[1], *Lu* Lu[1], *Junhong* Zhao[1], *Dali* Wang[2], *Wenlian* Lu[3] *and Jie* Yang[1*]

1 School of Mathematical Sciences, Dalian University of Technology, Dalian, Liaoning, China.
2 Central Laboratory, Dalian Children's Hospital of Dalian medical university, China
3 Institute of Science and Technology for Brain-inspired Intelligence, Fudan University

Abstract. CNN is an artificial neural network that can automatically extract features with relatively few parameters, which is the advantage of CNN in image classification tasks. The purpose of this paper is to propose a new algorithm to improve the classification performance of CNN by strengthening boundary samples. The samples with predicted values near the classification boundary are recorded as hard samples. In this algorithm, the errors of hard samples are added as a penalty term of the original loss function. Multi-classification and binary classification experiments were performed using the MNIST data set and three sub-data sets of CIFAR-10, respectively. The experimental results prove that the accuracy of the new algorithm is improved in both binary classification and multi-classification problems.

1 Introduction

Recently, various strategies have been introduced to boost the performance of Convolutional Neural Networks (CNNs). Irsoy et al. [1] proposed introduced dropout to prevent overfitting to increase the generalization. Zoumpourlis et al. [2] presented a second-order convolutional network that combines linear and non-linear filters. Juefei-Xu F et al. [3] introduced the local binary convolution (LBC) to replace the convolutional layer, which improved the computing time of the network. Uchida et al. [4] showed a couple convolutional layer with mutually constrained weights to produce better performance.

Some existing literatures optimize the learning algorithms for CNN. Zhining et al. [5] combined the genetic algorithm and CNN to extract better expression features than CNN dose. Hu et al. [6] adopted Gaussian convolution into CNN to accelerate the training convergence. Kim et al. [7] used the Extreme Learning Machine to calculate the weights between the hidden and output layers to shorten the training time. Yang et al. [8] combined the dropout and stochastic gradient descent optimizer to form a modified CNN algorithm, which improved the accuracy of CNN.

In this paper, a new learning algorithm based on strengthening boundary samples is proposed to improve the classification accuracy of CNN. The importance of each sample is adjusted in each training step according to its predicted value to pay more attention on the hard-classified samples. The rest of this paper is organized as follows. In Sect. 2, the structure of CNN is briefly introduced. In Sect. 3, we provide our new learning algorithm. Sect. 4 discusses the experimental results. Finally, Sect. 5 is the conclusion of this paper.

* Corresponding author: yangjiee@dlut.edu.cn

2 Related works

2.1 Convolutional neural network

Generally, a CNN contains several blocks including the convolution and pooling layers, and a fully connected layer. The convolutional layer has a number of convolution filters to extract the local characteristics of the image

$$x_j^l = f\left(\sum_{i\in M_j} x_i^{l-1} * W_{ij}^l + b_j^l\right) \tag{2.1}$$

where x_j^l denotes the j-th feature map in the l layer of the network, M_j is the set of feature maps in the l-1 layer, w_{ij}^l represents the convolution filter, b_j^l is the bias of the j-th feature map in the l layer, * is 2D convolution operation, and $f(\cdot)$ is the activation function. The pooling layer compresses data and parameters

$$y_j^l = \beta_j^l p(x_j^l) \tag{2.2}$$

where β_j^l is the weight and $p(\cdot)$ is the pooling function.

2.2 The new learning algorithm

The back propagation (BP) algorithm [9] is a popular algorithm to train CNN. Generally, the mean square error and cross-entropy loss are used as the loss function for two and multi classification problems, respectively.

The expressions of the two evaluation methods are as follows:

$$MSE = \frac{1}{n}\sum_{i=1}^{n}(t_i - y_i)^2, \quad (3.1)$$

$$CE = -\frac{1}{n}\sum_{i=1}^{n}t_i \cdot \log y_i, \quad (3.2)$$

where n is the batch size, t_i and y_i represent the target value and predicted value of the i-th sample, respectively.

It can be seen from equations (3.1) and (3.2) that all samples are equally important during the training process. In fact, some samples are classified with higher confidence, while others are classified with lower confidence. The network does not make reasonable adjustments to complex and simple samples during the training process, resulting in low classification performance. In this paper, we solve this problem by constantly changing the importance of training samples during the training process to appropriately strengthen the training for complex samples.

Firstly, we divide samples into hard samples and easy samples. Easy samples are higher or lower predicted values, because the effect of these samples and the corresponding loss are in line with our expectations. For example, the corresponding loss of a sample with a small value is large, therefore, it has a poor effect and a large impact in training, which is in line with our expectations. However, there are some samples whose prediction value are near the classification boundary value, and the effect on classification are very poor, but the corresponding loss values are relatively small, which will lead to lower classification accuracy. We call this part of the samples hard samples.

The specific distinguishing method is to take an interval near the classification boundary value, the samples with predicted values within the interval are recorded as hard samples, and the samples with predicted values outside the interval are recorded as easy samples. We use the average of the predicted values of the same batch of samples to select the interval. Take the binary classification as an example, let the classification boundary value be *0.5*, the average value be a, let the distance from a to *0.5* be d, take *0.5* as the center value of the interval, and *2*d* as the interval length to obtain the interval [*0.5-d, 0.5 + d*]. Figure 1 shows the sample division in the binary classifications.

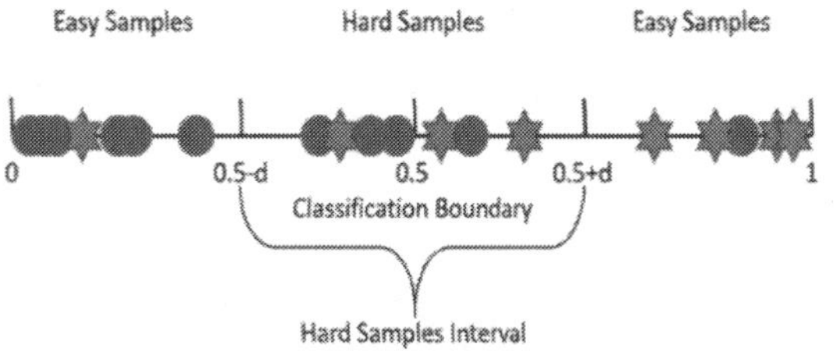

Fig. 1. The Classification of sample

Then we use a new algorithm, PCNN, to strength learning hard samples by adding a penalty term to the loss function. The penalty term extracts the information of the degree to which the sample is classified difficultly. The loss function for PCNN is expressed as:

$$NMSE = \frac{1}{n}\sum_{i=1}^{n}(t_i - y_i)^2 + \lambda\left[\frac{1}{m}\sum_{k=1}^{m}(y_k^{'} - (0.5 - d_k))^2 + \frac{1}{t}\sum_{l=1}^{t}((0.5 + d_l) - y_l^{''})^2\right] \quad (3.3)$$

$$NCE = -\frac{1}{n}\sum_{i=1}^{n} t_i \cdot log\, y_i - \lambda\left[\frac{1}{m}\sum_{k=1}^{m}(log\, y_k^{'} - log(0.5 - d_k)) + \frac{1}{t}\sum_{l=1}^{t}(log(0.5 + d_l) - log\, y_l^{''})\right] \quad (3.4)$$

where Equation (3.3) is the mean square error of the PCNN, Equation (3.4) is the cross-entropy loss function of the PCNN, λ is the proportion of the penalty term in the loss function, and the specific value of λ is determined by subsequent experiments, m and t are the number of hard samples on the left and right of *0.5* in the batch, respectively, $y_k^{'}$ is the value of the k-th hard samples to the left of *0.5*, $y_l^{''}$ is the value of the l-th hard samples to the right of *0.5*, and d_k, d_l are the average of the predicted values of the k-th and the l-th hard sample categories in the batch, respectively.

When $\lambda = 0$, the new loss function is equal to the original loss function. Therefore, the modified loss function can be regarded as a generalization of the original loss function. In addition, the effect of the penalty term changes as the value changes. By choosing the value of λ, the CNN can be better trained and avoid overfitting. In the experiment, since the accuracy is highest when the value of the penalty term $\lambda = 0.7$, $\lambda = 0.7$is selected for subsequent experiments. Figure 2 is the corresponding change of the classification accuracy with respect to the value of λ on the MNIST data set.

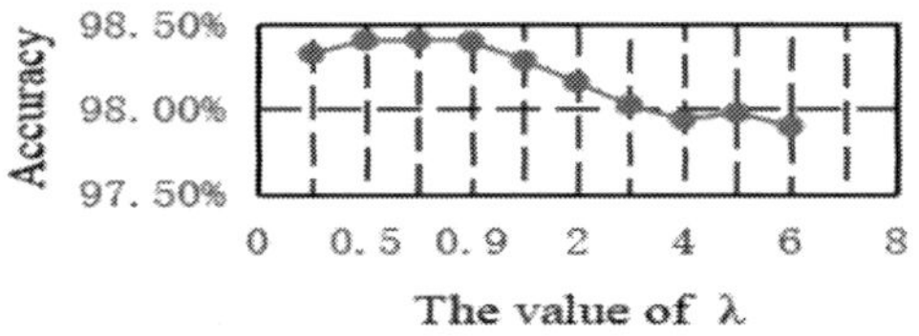

Fig. 2. Classification accuracy varies with λ

4 Experiments

In this section, some experiments are performed to test the effectiveness of the new learning algorithm for CNN on CIFAR-10 and MNIST datasets. The architecture of CNN is designed as shown in table 1. The ReLU function is used as the activation function [13] and Dropout [14] is used to avoid overfitting. In all experiments, the learning rate is set as 0.001 and the batch size choose 50. 10 experiments with different initial weights are performed on each dataset and the average accuracies are recorded.

Table 1. The architecture of CNN.

Layer	CIFAR-10	MNIST
Input layer	32×32×3	28×28×1
Conv layer	3×3×64	3×3×64
Conv layer	3×3×64	3×3×64
Max pooling layer	2×2	2×2
Conv layer	3×3×128	3×3×128
Conv layer	3×3×128	3×3×128
Max pooling layer	2×2	2×2
Conv layer	3×3×256	3×3×256
Conv layer	3×3×256	3×3×256
Conv layer	3×3×256	3×3×256
Fully Connected layer (dropout)	1024	1024
Output layer	1	10

4.1. CIFAR-10

The pictures in the CIFAR-10 are equally divided into ten categories and each category contains 5,000 training samples and 1,000 test samples. We constructed three sub-data sets for the binary classification problem from the CIFAR-10 dataset, of which the first and the second types constitute the first sub-data set, the fifth and the eighth to the second sub-dataset, and ninth and tenth to the third sub-dataset. In these sub-datasets, the loss function is Equation (3.3).

Figure 3 shows the test accuracies of PCNN and CNN for the first sub-dataset in different iterations. As we can see that PCNN perform better than CNN at any time. When iterations=1600, the accuracy of CNN reached 95.495%, while the accuracy of PCNN reached 95.885%. And in the whole process, the performance of PCNN is better than CNN, especially when the number of epochs is less than 800. If the number of iterations is 160, the classification accuracy of PCNN is 3.37% higher than that of CNN.

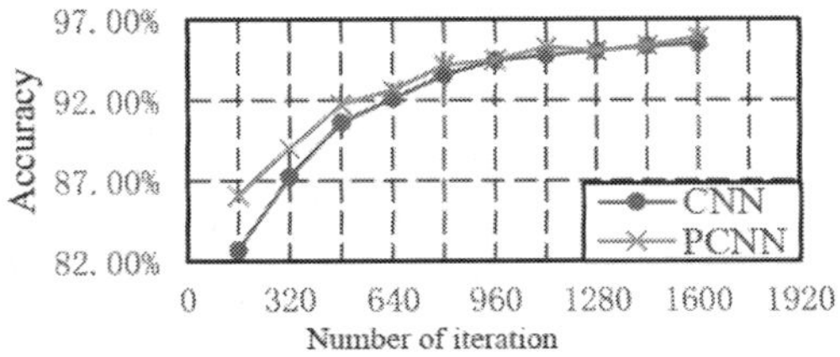

Fig. 3. Comparison on test set between CNN and PCNN on the first sub-dataset.

On the second sub-dataset, Figure 4 shows the classification accuracy curves of PCNN and CNN under different iterations. We can see that the PCNN curve is always higher than the CNN. With iteration=1600, the classification accuracy of traditional CNN is 86.95%, and the classification accuracy of PCNN is 87.306%. If the number of iterations is 640, the classification accuracy of PCNN is 1.188% higher than that of CNN.

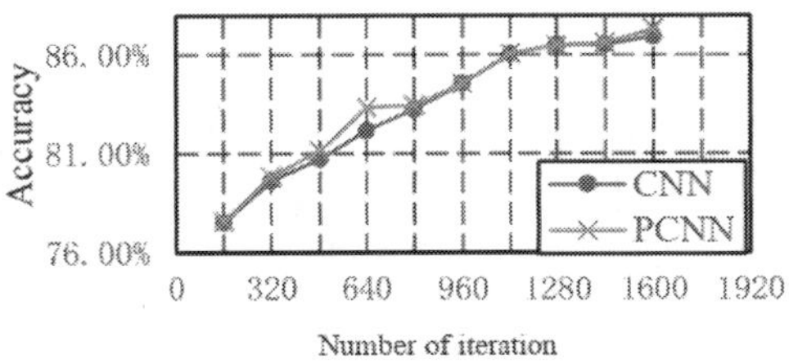

Fig. 4. Comparison on test set between CNN and PCNN on the second sub-dataset.

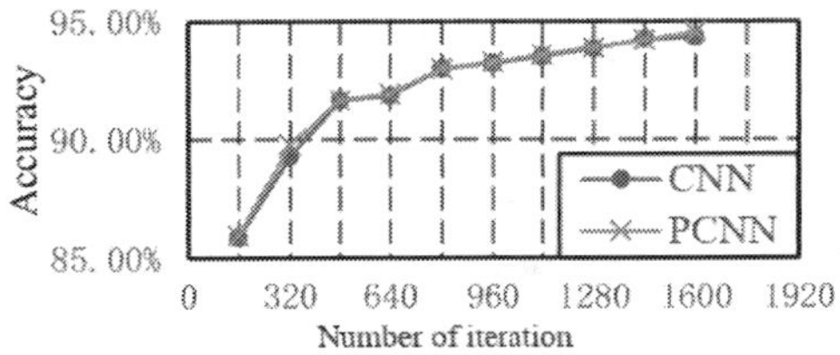

Fig. 5. Comparison on test set between CNN and PCNN on the third sub-dataset.

Figure 5 shows the classification accuracy curves of PCNN and CNN on the third sub-dataset. It is obvious that whether on PCNN or CNN, the classification accuracy improves with the increase of the times. After 1600 iterations, the accuracy of CNN reached 94.3%, and the accuracy of PCNN reached 94.54%. In this dataset, the effect of PCNN is not very obvious. This is because the difference between the hard and simple samples is small and the effect of the penalty term is small.

4.2. MNIST

MNIST is the dataset consists of handwritten digital pictures, which is divided into 10 categories, and the corresponding labels are 0 to 9 respectively. In this experiment, we use Equation (3.4) as a loss function to explore the impact of PCNN on multi-classification problems.

Experiments are performed on PCNN and CNN with different iterations, and the results are shown in Figure 6. It can be seen that the classification accuracy of PCNN can always be higher than CNN. Specifically, when the number of epochs is less than 11,000, the accuracy of PCNN is significantly higher than that of CNN. And when the number of periods is greater than 11,000, the improvement of PCNN becomes smaller. The reason for this phenomenon is that with time, the prediction accuracy of all samples will increase, and the increased error will decrease. After 33,000 iterations, the accuracy of CNN and PCNN is 99.165% and 99.185%, respectively.

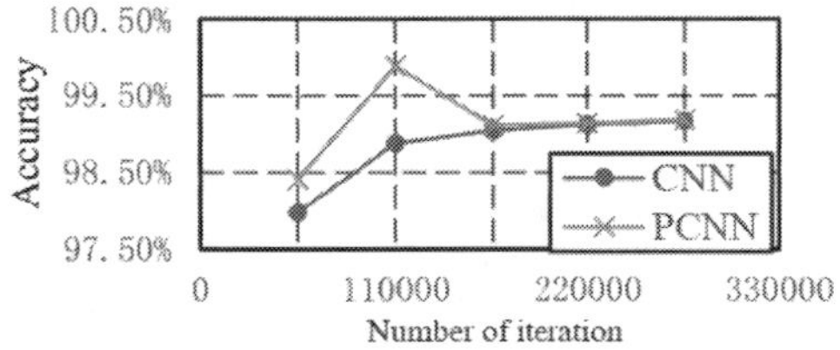

Fig. 6. Comparison on test set between CNN and PCNN on MNIST dataset.

5 Conclusion

In this paper, a new learning algorithm is proposed to enhance the classification performance of CNN. It focuses on dividing samples into easy and hard samples to strength learning the hard samples by adding a penalty term to the loss function of BP. Experimental results on the CIFAR-10 and MNIST datasets show that the new learning algorithm can improve the classification accuracy and speed of CNNs in image classification problems. In future study, we will try to set the weight of the penalty term with a dynamic parameter to make it adaptively to improve the performance of CNN.

The authors would like to thank the editor and reviewers for their constructive comments that greatly improved the quality of this paper. This work was supported by National Key R&D Program of China (No.2018AAA0100300), Shanghai Municipal Science and Technology Major Project (No. 2018SHZDZX01) and ZJLab and the National Natural Science Foundation of China (No. 11201051).

References

1. Irsoy O and Alpaydinodotn E (2018) Dropout Regularization in Hierarchical Mixture of Experts *arXiv* **1812.10158**
2. Zoumpourli G, Doumanoglou A and Vretos N,et al. (2017) Non-linear Convolution Filters for CNN-based Learning *IEEE International Conference on Computer Vision* **4771-4779**
3. Juefei-Xu F, Boddeti V and Savvides M (2017) Local Binary Convolutional Neural Networks *IEEE Conference on Computer Vision and Pattern Recognition* **4284-4293**
4. Uchida K, Tanaka M and Okutomi M (2016) Coupled Convolution Layer for Convolutional Neural Network *International Conference on Pattern Recognition* **3548-3553**
5. Zhining Y and Yunming P (2015) The genetic convolutional neural network model based on random sample Int. Journal of u-and e-Service, *Science and Technology* **8 pp 317-326**
6. Kim J, Kim J and Jang G J (2017) Fast learning method for convolutional neural networks using extreme learning machine and its application to lane detection Neural Networks **87 109-121**
7. LeCun Y, Boser B E and Denker J S (1990) Handwritten digit recognition with a back-propagation network *NIPS (Denver)* **396-404**
8. Nair V and Hinton G E (2010) Rectified linear units improve restricted boltzmann machines *ICML-10 (Haifa)* **807-814**
9. Srivastava N, Hinton G and Krizhevsky A (2014) Dropout: a simple way to prevent neural networks from overfitting *The Journal of Machine Learning Research* **15 1929-1958**

Improvement of high temperature fault problem of Wind power inverter based on TRIZ

LinLin Sun[1]*; Yang Liu*[2]*; JiaYu Xie*[1] *; Rui Cao*[1]

[1]Jilin University, College of Mechanical and Aerospace Engineering, Changchun 130022, China
[2]Datang Jilin Power Generation Co, Ltd.130022, China

Abstract. The research content of this paper is to improve the high temperature fault problem of the wind power inverter and optimize the processing process of the high temperature fault problem. In order to solve the high temperature fault problem, the water-cooling system of inverter is analyzed in detail by using the function analysis, cause and effect analysis and object field analysis in the invention problem solving theory (TRIZ theory). At the same time, combined with the invention principle of TRIZ theory and the standard solution, seven different solutions are finally obtained. Based on the ECRS principle, the general process of high temperature fault problem treatment is simplified through continuous optimization and improvement.

1 Introduction

As an important part of the wind power plant connected with the power grid, inverter has considerable social benefits and a wide range of applications. However, the inverter will generate a lot of heat in the process of operation, which will lead to frequent high temperature failures, and even cause serious accidents and economic consequences.

At home and abroad, there are many research achievements on inverter heat dissipation. Foreign research on liquid cooling related technologies started very early, and related technologies are becoming mature. Although China started relatively late and still lags behind the international advanced level in terms of core components, many new technological achievements have emerged in recent years with rapid development and good momentum. At present, foreign research has been carried out on the material transformation of cold plate passage. Porous material heat sink is used as the material to improve its heat transfer efficiency [1]. The fin of the fluid passage is created, and the radiator is improved obviously and intuitively [2]. By modifying the internal structure of the cold plate, the triangular fluid path is obtained to improve its heat transfer effect [3]. In recent years, a new equivalent thermal circuit diagram considering NTC thermistor in IGBT module has been proposed in China [4]. The cooling device of the megawatt power device IGBT was optimized, and the cold plate structure with superior comprehensive performance was obtained [5].

Although the research on inverter is relatively mature, there are few inverter heat dissipation achievements and flow optimization in the field of wind power generation. Therefore, based on the TRIZ theory and ECRS principle, this paper will improve the high temperature fault and the cooling system of the inverter and optimize the process of fault handling.

2 Current situation and problem analysis of inverter and water cooling system

2.1. Inverter with water cooling system

2.1.1 Net side inverter

The grid-side inverter module is located in the variable-flow control cabinet, and the ceramic substrate of its internal IGBT module is connected with the water-cooled substrate via the heat-conducting silica gel. There is a wind cooling system in the converter control cabinet, which can create a wind cooling cycle in the converter control cabinet to avoid local overheating of the inverter.

2.1.2 Water circulation cooling system

(1) Cooling control cabinet

The water circulation cooling system is located in the cooling control cabinet. The cooling control cabinet consists of two parts: the circulating cooling system and the measuring control system. The schematic diagram of the water circulation cooling system is shown in FIG. 1.

* Corresponding author: slfdly6009@163.com

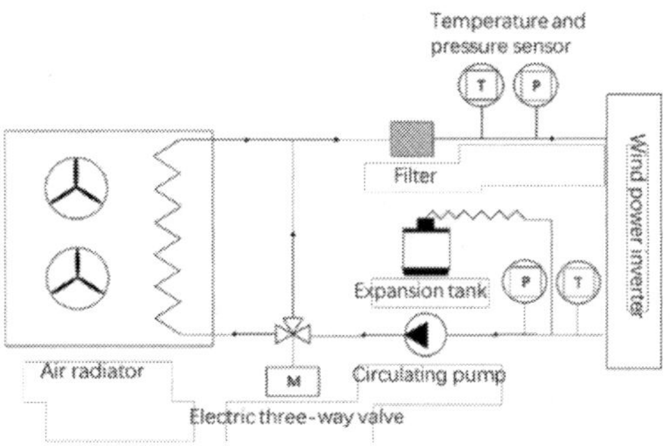

Fig. 1. Schematic diagram of water circulation cooling system

(2) PLC

The core of the measurement control system in the cooling control cabinet is PLC, which monitors the cooling circulation system in real time.

(3) Cooling fluid

The coolant is a mixture of purified water, ethylene glycol and preservatives.

(4) Water cold plate pipe

The water enters the water-cooled substrate from the inlet pipe, then flows through the water-cooled pipe, and then out of the outlet pipe, where the water passes right across the bottom of the IGBT.

(5) Water cooling system parameters

See Table 1. for the parameter setting of water cooling system:

Table 1. Water cooling system parameter setting

The serial number	Project	Normal values
1	System static pressure	2.0bar
2	Dynamic pressure of inlet valve	2.1bar
3	Exit valve dynamic pressure	1.7bar
4	Flow	290L/min
5	Inlet valve temperature	<45°C
6	Temperature difference of inlet and outlet valve	<5.5°C

2.2 Existing problems

With the continuous increase of the ambient temperature, when the wind power generating unit reaches full capacity in the windy weather, "grid side inverter temperature fault" is frequently reported. Table2. and Table3. are the fault file data of different locations(Both Bayin and Guantin are local place names), as shown below:

Table 2. Bayin scene field failure file data (Temperature of the unit/C°)

Machine number	Ambient temperature	Inlet valve temperature	Outlet valve temperature	Inverter temperature	Fault time
38#	24.5	40.5	42.9	68	13：09
51#	32.6	38.1	40.1	68	14：15
63#	23.8	37.1	39.1	68	15：51
89#	24.3	37.3	39.4	68	21：54
89#	23.8	37.7	39.9	68	23：43
95#	24.7	38.3	40.4	68	15：12
95#	25.0	38.1	40.1	68	16：18

Table 3. Guantin scene field failure file data (Temperature of the unit/C°)

Machine number	Ambient temperature	Inlet valve temperature	Outlet valve temperature	Inverter temperature	Fault time
13#	34.0	40.9	42.5	68	13：10
15#	35.6	40.6	42.3	68	15：11
16#	32.3	38.3	40.0	68	12：16
27#	32.3	40.9	42.9	68	12：10
3#	35.3	40.5	42.2	68	15：10
35#	34.0	40.4	42.2	68	13：05
87#	35.2	42.5	43.7	68	15：17

According to the above table, the temperature difference between the on-site inverter temperature and the outlet valve temperature is about 26 ℃, which is obviously too high, indicating that the heat dissipation efficiency between the inverter IGBT module and the water cooling system is not enough. It can be seen that the water cooling system needs to be improved, and the specific analysis is as follows:

The difference between Bayin scene and Guanting scene fault file data is the temperature difference between the inlet valve water temperature and the ambient temperature.The temperature difference on site in Guanting is about 7 ℃, which is 2 ℃ higher than the 5 ℃ required by the design, indicating that there is still a certain lack of radiator cooling capacity, which did not fully meet the requirements of the radiator design at that time.The temperature difference in Bayin is about 14 degrees Celsius, more than twice the original radiator design requirements.

There are two main reasons for this large discrepancy: One is the natural wind,At that time, the wind direction of Bayin site was just opposite to the blowing direction of radiator fan, so the radiator cooling effect was reduced.The wind direction at the time of the unit failure in guanting was exactly 90 degrees to the blowing direction of the radiator fan, so it had no significant impact on the radiator's cooling effect; second, the radiator did have insufficient cooling capacity.

To verify this, the measured data from Bayin site are shown in Table 4 (the wind direction is exactly 90 degrees from the direction of the radiator fan at the time of measurement).

From the measured data can be seen that the temperature difference between the inlet valve water temperature and the ambient temperature is about 9 degrees Celsius, compared with the original 14 degrees Celsius has a great improvement, so it can be explained that the wind direction of the radiator heat dissipation will also have a great impact.

Table 4. Measurement data of Bayin site（Temperature of the unit/C°）

Machine number	Ambient temperature	Inlet valve temperature	Outlet valve temperature	Inverter temperature	Radiator temperature range
1#	25.5	38.5	39.7	63	30—37
2#	25.6	33.6	35.1	61	26—34
10#	26.1	36.5	38.1	61	29—36
19#	26.5	34.1	35.8	59	29—34
8#	26.7	34.3	35.5	60	27—34
25#	26.8	35.0	36.1	63	24—33

3 Analysis and improvement of high temperature fault based on TRIZ

3.1. Functional analysis

3.1.1 Building functional models

Combining the interaction matrix and the function list of components, the function structure model is established, as shown in the fig 2:

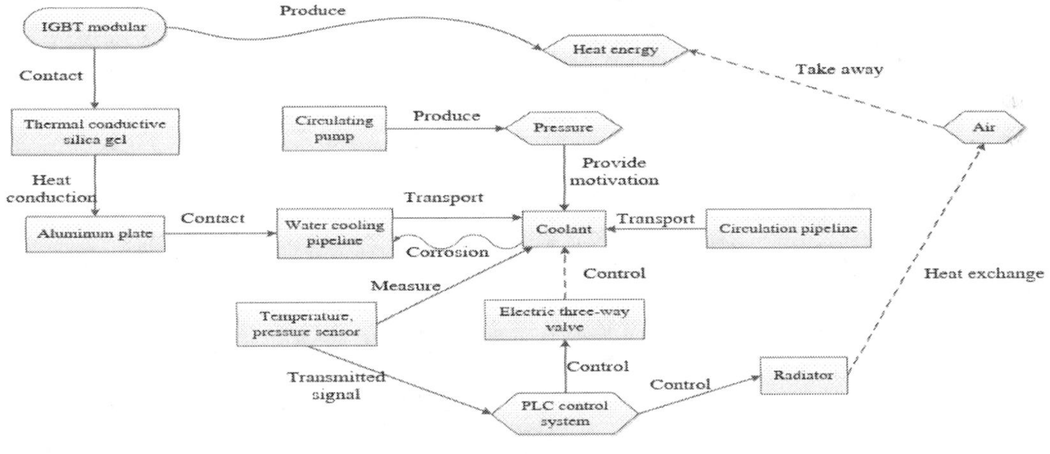

Fig. 2. Functional structure model

3.1.2 Field analysis and solution

(1) For the functional analysis of water-cooling pipes and coolants, S1 = water-cooling pipes, S2 = coolants.

The function model is represented by symbol system as shown in Fig 3:

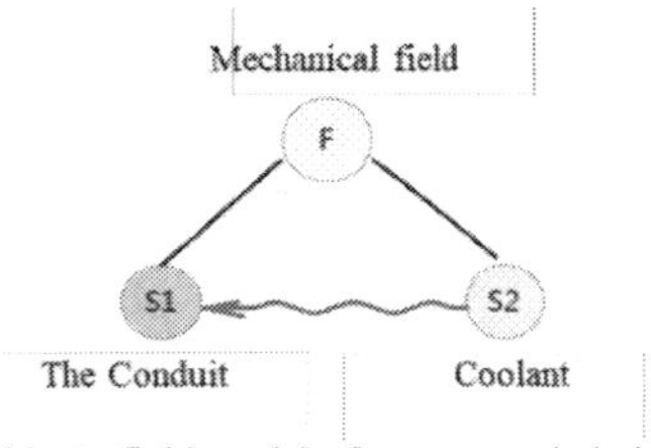

Fig. 3. Object - field model of water - cooled pipeline and coolant

Using standard solution 1.2.2 to introduce improved S1 or (and) S2 to eliminate harmful effects, that is, the existing system is complete, harmful and useful effects exist at the same time, the third substance S3 can be introduced between substances to eliminate harmful effects, S3 is the deformation of existing substances.
Scheme 1 is proposed: : S3 is the deformation of S2 coolant. The existing coolant is composed of glycol and purified water, which is corrosive and easy to cause harmful effects on the pipeline. The introduction of 1,2-propanediol, which is non corrosive and easy to degrade, can reduce the corrosion effect of coolant on metal pipes.
(2) For the functional analysis of the electric three-way valve and coolant, S1 = coolant, S2 = electric three-way valve. The function model is represented by the symbol system as shown in Fig 4.

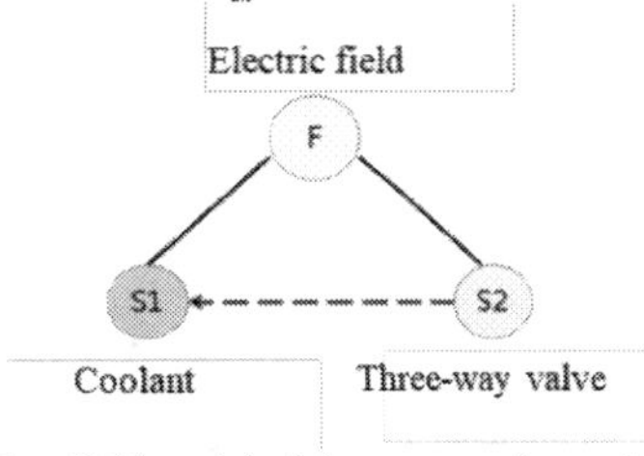

Fig. 4. object field model of three-way valve and coolant

Using more controllable field by standard solution 2.2.1: the controllability of the object field model is insufficient. The field that is easy to control can be used to replace the field that is difficult to control (from difficult to easy, it can be arranged as mechanical field, thermal field, chemical field, electric field and magnetic field), so as to improve the controllability.

Scheme 2 is proposed: : magnetic control valve is used instead of electric three-way valve to achieve more accurate control.
(3) For the functional analysis of cooling fan and air, S1 = air, S2 = radiator. The function model is represented by symbol system as shown in Fig 5:

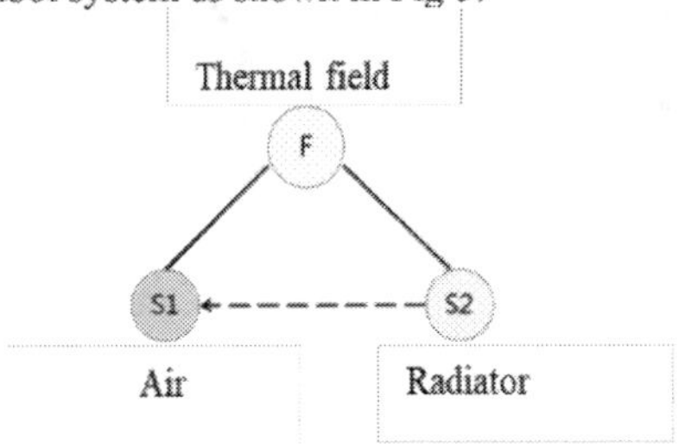

Fig. 5. object field model of cooling fan and air

Use standard solution 2.2.3 to use substances with capillary and porous structure: change the structure of substances to become substances with capillary or porous structure, and let gas or liquid pass through these substances with capillary or porous structure, so as to obtain the enhancement of system function effect.
Scheme 3 is proposed: change the fins in the radiator into fins with staggered hole structure, increase the heat exchange area, so as to improve the heat exchange efficiency of the radiator.

3.2. Causal analysis

3.2.1 Causal analysis diagram

The causal analysis diagram is shown in fig 6:

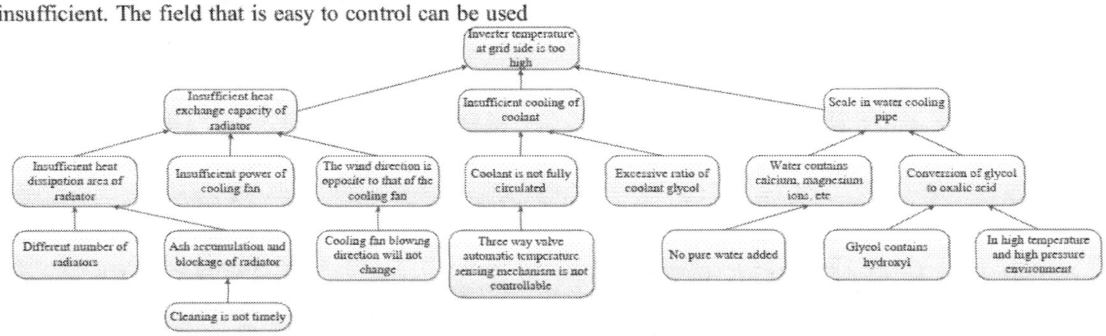

Fig. 6. Cause and effect analysis chart

Considering the practical problems such as optimization cost and feasibility, four root causes are summarized: (1) the radiator is blocked by ash deposition; (2) the wind direction of the cooling fan does not change; (3) the automatic temperature sensing mechanism of the three-way valve is not controllable; (4) the content of glycol in the coolant is not accurate. Through 40 principles of invention recommended by contradiction matrix, the causes of the four root terms are analyzed.

3.2.2 Scheme analysis

(1) The radiator plate is easy to be blocked by ash accumulation. Generally, it is cleaned regularly by manpower with high-pressure water gun, which requires

more manpower and time cost, as well as equipment shutdown operation. Therefore, there is a technical contradiction. The improved parameter is the harmful factor of the object, and the deteriorated parameter is the time loss. The principles of invention 35, change of physical or chemical parameters, 18, mechanical vibration, 34 abandonment and regeneration are applied.
Inspired by the change principle of physical or chemical parameters, scheme 4 is proposed: the radiator with self-cleaning function is obtained by coating a layer of nano TiO film on the surface of the radiator.

(2) When the wind direction is opposite to the blowing direction of the cooling fan, the heat transfer capacity of the radiator will be affected, and the wind direction is uncontrollable. Therefore, it can be considered to change the direction of the cooling fan, so that the radiator can be adjusted to rotate 90 degrees. But adding a device that can adjust the direction of radiator increases the complexity of the device. At this time, the parameters of technical contradiction improvement are adaptability and universality, and the parameters of deterioration are the complexity of the system. According to the contradiction matrix, the principle of invention is 15 dynamic characteristic principle, 34 abandonment and regeneration principle, 1 division principle, 16 unreachable or excessive action principle.
According to the principle of dynamic characteristics and the principle of partition, scheme 5: improve the dispersion degree of air radiator and water cooling system, and make it supported by the load-bearing device which can adjust the direction. And a wind direction sensor is set, which can make the device adjust the angle with the change of wind direction.
(3) The automatic temperature sensing mechanism in the three-way valve regulates the amount of internal and external circulation of coolant according to different water temperature, so it can not control the internal and external circulation of coolant arbitrarily according to the actual needs. Want to get more accurate and convenient control, but at the same time change its automation. That is to say, the improved parameter is the convenience of the operation process, and the deteriorated parameter is the degree of automation. The invention principle is as follows: 1. Division principle, 34. Abandonment and regeneration principle, 12. Equipotential principle, 3. Local mass principle.
By using the segmentation principle, scheme 6 is obtained: setting two modes: automatic mode and manual mode. In the manual mode, the coolant flow rate of inner and outer circulation can be determined according to the actual situation, which is more convenient to control.

(4) When the coolant is used for antifreeze, it is required to increase the content of glycol in the coolant; when it is necessary to play the cooling and heat dissipation role of the coolant, it is necessary to reduce the content of glycol in the coolant. There is a physical contradiction.
Using time separation, we can get scheme 7: in the period of frequent high temperature failure, we can take out about 10% of the coolant in the water cooling system, and then add 10% of the pure water into the coolant. In winter, about 10% of the coolant in the water cooling system can be taken out, and then 10% glycol can be added to the coolant for replenishment.

4 Process optimization of high temperature fault treatment based on ECRs principle

4.1. High temperature fault

4.1.1 Fault description

Fan fault is usually found by fan alarm and artificial fan data analysis.

Table 5. Explanation of inverter high temperature fault

Category	Explain
Fault name	High temperature of grid side inverter
Fault description	Temperature over limit
Failure time	100ms
Fault set value	67°C
Reset value	57°C

4.1.2 Cooling system related faults

Although the small problems of each component of the cooling system are less than the degree of failure, they will also affect the heat exchange efficiency, thus affecting the temperature of the inverter. There are 13 related problems, such as temperature fault of control cabinet, fan fault of water cooling system, pressure fault of inlet and outlet valve, temperature fault of inlet and outlet valve, low pressure fault of inlet valve, warning of water cooling fan, etc.

4.2. General processing flow of inverter high temperature fault

The general processing flow chart of inverter high temperature fault is shown in Fig 7:

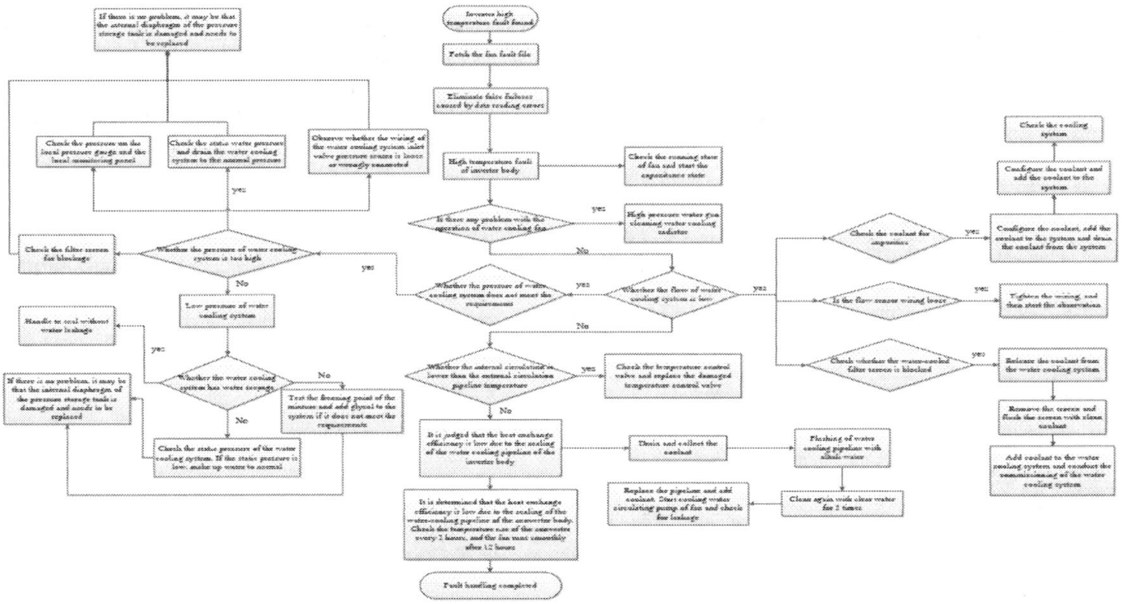

Fig.7. general processing flow chart of inverter high temperature fault

4.3. Use ECRS principle to optimize the process

4.3.1 Cancel

(1) When there is a problem in the air cooling system of the converter control cabinet, the possibility of high temperature fault reported by the converter module and the controller module at the motor side is greatly increased. If only the inverter module reports high temperature fault, the temperature of the controller and converter is within the normal range, and there is no sudden rising trend of temperature, the step of checking the cooling fan of the control cabinet can be cancelled. the two steps of checking whether the flow sensor wiring is loose and whether the temperature sensor is loose can also be cancelled.

(2) Because the staff will regularly check the operation of the cooling fan and clean it. Therefore, if a high temperature fault occurs within a short time after the inspection and cleaning time, it can be considered to cancel the inspection of this step.

(3) The real-time monitoring data of the main controller shows that the pressure of the water-cooling system is on the high side, while the pressure on the local pressure gauge and the monitoring panel is normal, which indicates that the pressure sensor of the inlet valve has problems, transmits error signals, and the pressure of the water-cooling system is normal. It shows that this detection step has little correlation with high temperature fault, and can be cancelled. Similarly,

(4) When the pressure of the water cooling system is low, it may be due to the ice phenomenon of the mixture in the low temperature environment. Therefore, when the pressure of the water cooling system is low, there will be the ice point value of the test coolant. If it does not meet the requirements, add glycol to the coolant in the water cooling system, so as to reduce the possibility of ice formation at low temperature. But because this is to deal with the inverter high temperature fault, the ambient temperature of high temperature fault will not be low, so this step is canceled.

4.3.2 Merge

The low flow of water-cooling system is caused by abnormal system pressure or clogged filter screen. The abnormal system pressure may also be caused by clogged filter screen. Therefore, there are two times to check whether the filter screen is clogged in the process. Therefore, the filter screen shall be merged and replaced until the low flow of water-cooling system is found.

4.3.3 Rearrangement

According to 30 seconds data before and after the fault to eliminate the false fault caused by data reading error, at the same time, judge whether the flow of water-cooling system is low and whether the pressure of water-cooling system meets the requirements. Comparing the flow of water-cooling system with the normal value in advance to eliminate the false fault caused by data reading error, it is more convenient and time-saving to avoid multiple data viewing at different times.

4.3.4 Simplify

(1) When the filter screen of the water-cooling system is blocked, the filter screen needs to be removed from the water-cooling system for cleaning. Two ball valves can

be installed on both sides of the filter screen. When cleaning is needed, the ball valve can be closed, so it is not necessary to discharge all the coolant to supplement these two steps, which simplifies the operation, saves a lot of time, and improves the cleaning efficiency of the filter screen.

(2) When the scale accumulation in the water-cooling pipeline is washed with the alkali water at the ratio, the reverse flushing method can be adopted, that is, the single-module outlet pipe of the original inverter is connected with the outlet pipe of the water-adding device, and the inlet pipe of the original single-module inverter is connected with the water tank of the water-adding device, so the flushing efficiency will be higher and the flushing time can be shortened.

4.3.5 The optimized process

The processing flow of high temperature fault of the optimized inverter is shown in Fig 8:

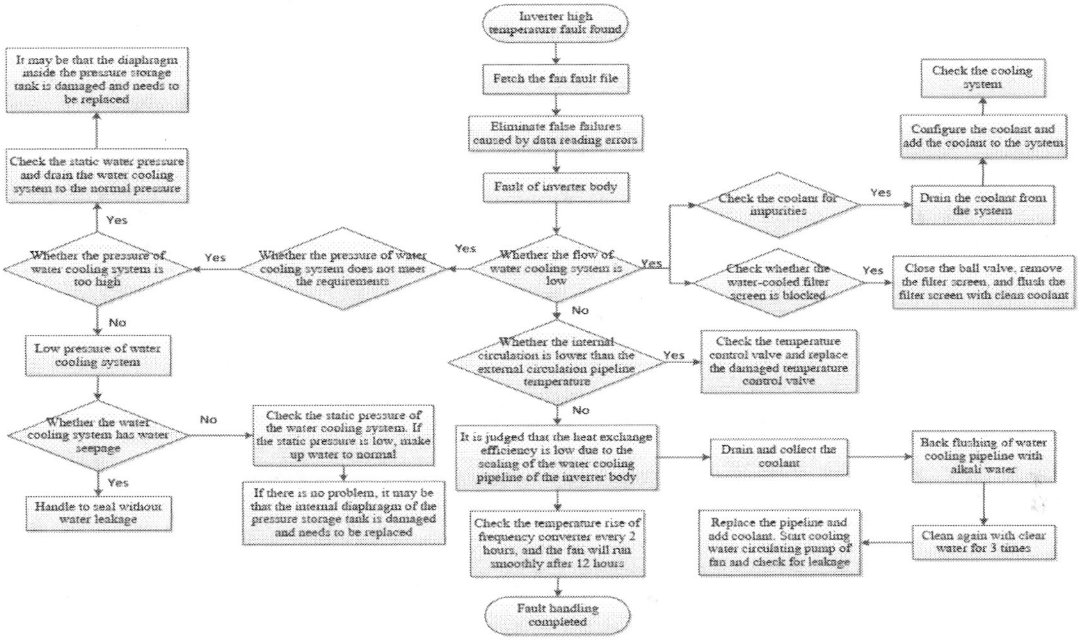

Fig.8. Processing flow of high temperature fault of inverter after optimization

5 Conclusions

By using TRIZ theory, this paper analyzes the high temperature fault of wind power inverter and the improvement of cooling system, and puts forward a feasible scheme. Combined with the functional analysis, the cooling system is introduced into the non corrosive and easy to degrade 1, There are three schemes: 2-propanediol, magnetic control valve instead of electric three-way valve, changing the radiator fins into fins with staggered hole structure; combined with cause and effect analysis, four schemes are obtained: changing the surface material of radiator fins, designing a wind direction sensor, setting manual and automatic modes, changing the content of glycol in the coolant timely. It shows that TRIZ theory provides people with an efficient and innovative way of thinking and plays a strong role in the actual industrial production process. At the same time, through the principle of ECRs, the process of troubleshooting is optimized, the processing cost and time are reduced, and the production efficiency is increased.

References

1. Lu G, Zhao J, Lin L, et al. A new scheme for reducing pressure drop and thermal resistance simultaneously in microchannel heat sinks with wavy porous fins[J]. International Journal of Heat and Mass Transfer, 111: 1071-1078.(2017)
2. Kanargi O B, Lee P S, Yap C. A numerical and experimental investigation of heattransfer and fluid flow characteristics of a cross-connected alternating converging-diverging channel heat sink[J]. International Journal of Heat and Mass Transfer ,106:449-464. (2016)
3. Wong K, Lee J. Investigation of thermal preformance of microchannel heat sink with triangle ribs in the transverse microchambers[J]. International Communications in Heat and Mass Transfer, 65: 103-110. (2015)
4. HongLu Zhao,YongYuan Zhu,Yin Zhang.Heat dissipation design of high power inverter[J].Electrical technology, 19(08):149-156. (2018)
5. ShuYing Tang. Cold plate structure optimization design of megawatt wind power converter[D]. Zhejiang University, (2016).

Accelerated testing of track vehicle torsion bar

Jan Furch[1,*], *Quy Hung* Nguyen[2], *Jiri* Stodola[1]

[1]University of Defence in Brno, Faculty of Military Technology, 61200 Brno, Czech Republic
[2] Institute of Mechanical Engineering, 122100 Hanoi, Vietnam

Abstract. The suspension system of track vehicles is often equipped with torsion bars. The torsion bar is a key element of the vehicle's suspension system. The service life of torsion bars depends primarily on the quality of the material used, the diameter of the torsion bars and the method of attachment and placement. In order to obtain the characteristics of the torsion bar, an experimental test stand was created on which accelerated tests were performed. Furthermore, a virtual model of torsion bar operation was created. The torsion bar was modelled using finite elements (FE part). Dynamic simulation was performed using MSC Adams software. Finally, the values measured in the accelerated tests and the results of simulations were compared. Comparing the values found, it was proved that this difference is not significant. This virtual model can be used to investigate the service life of torsion bars and to predict their service life.

1 Introduction

When test acceleration is accomplished with a single stress only, the models are life stress models, where the damage per unit time of the test is appropriately accelerated by increasing the level of stress.

The three most frequently used relationships are [1]:

a) inverse power law model, used for test acceleration when stresses other than constant temperature are considered, such as electrical, mechanical, chemical (corrosion) and others;
b) Arrhenius reaction rate model, used for constant temperature stresses, based on the effect that the absolute temperature has on a failure mechanism;
c) Eyring model which is used in cases where the acceleration is achieved with temperature and moisture stress levels. The model is derived from quantum mechanics.

With all acceleration models, test data can be analyzed using established analytical models to determine characteristic accelerated life parameters. Using the acceleration factors, the parameters corresponding to use environments are determined and used for reliability projections as needed. The acceleration models should, if possible, be verified by plotting the test data.

2 Planning lifetime tests

Planning lifetime tests can be divided into experimental-technical measurement planning and statistical test planning [2].

2.1 Experimental-technical measurement planning

The common fundamental principles for correct execution of an experiment apply. The most important of these principles is as listed the inverse power law is applicable to:

- The boundary conditions and limits must be exactly defined and kept.
- For lifetime tests this is especially important for the load spectrum.
- The technical measurement process for the registration and control of the boundary conditions must be established along with their accuracy. Depending upon the resources, more information is acquired at the test stand than actually needed.
- If longer testing times are expected, then the use of automated and/or computer controlled measured value gathering and control equipment should be strived for.
- For a determination of the lifetime, the exact specification of a limit value is necessary, at which the nominal function is no longer fulfilled. If the damage is a continuously changing value, as for example a leak volume for a seal.
- The control equipment must be built up in such a way that the primary failure cause can even be determined after the failure effect. This is important since each failure mode is assigned its own characteristic reliability parameters [2].

2.2 Statistical test planning

Accelerated life tests are component life tests with components operated at high stresses and failure data observed. Test planning means picking stress levels and sample sizes and test times to produce enough data to fit models and make projections. It is good design practice to put more of your test units in the lower stress cells, to make up for the fact that these cells will have a smaller proportion of units failing.

* Corresponding author: jan.furch@unob.cz

In statistical test planning, the first step involves determining the size of the inspection lot. The inspection lot size is in close connection with the confidence levels and the statistical spread of the measured values. If fewer components are tested, then the result of the statistical assessment becomes more uncertain. For an accurate result, it is necessary that a sufficient quantity of components is tested. This can increase the time and effort involved in a test immensely. Fig. 1 shows the influence of number of test units on the reliability. It is obvious that with the diminishing of sample size, the reliability is decreasing.

Another important point in statistical test planning involves establishing a suitable test strategy. Possible strategies include:

a) complete tests,
b) incomplete (censored) tests,
c) strategies for shortening test times.

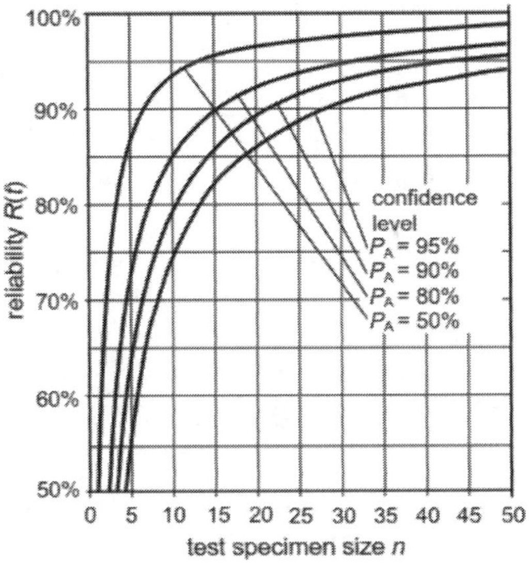

Fig. 1. Minimal reliability $R(t)$ as a function of the test specimen size n and the confidence level P_A, if at the point in time t no failure has occurred [2]

Another option for a considerably shorter test time is the Sudden Death Test and tests with an increased load. There are different types of accelerated life test (ALT) plans in use, which include subjective, traditional, best traditional, statistically optimum and compromise plans [3], [4], [5]. As with the simpler ALTs, evaluation of test plan properties can be done using either large-sample approximations or simulation methods [3].

Monte Carlo simulation provides a powerful, insightful tool for planning experiments. For a specified model and planning values for the model parameters, it is possible to use a computer to simulate ALT experiments to see the kind of data that will be obtained and to visualize the variability from trial to trial. Such simulations provide an assessment of sampling uncertainty that will result from using a limited number of test specimens [6].

3 Theory planning accelerated life test

Before starting an accelerated life test, it is advisable to have a plan that helps in accurately estimating reliability at operating conditions while minimizing test time and costs. To design the plan for the accelerated life tests it is necessary to establish the following parameters:

1 The acceleration model: For accelerated life tests where the failure mechanism is a mechanical one (the fatigue failure), the most adequate one is the IPL model. The inverse power law (IPL) model is commonly used for non-thermal accelerated stresses. This model was calculated by the formulas (1) - (3).
With the inverse power law, the characteristic that represents product reliability related to time, such as characteristic life, mean life, mean time to a failure, is represented as [1]:

$$L(S) = C^{-1} \times S^{-m} \quad (1)$$

where: $L(S)$ = is the life or other predetermined time duration as a function of stress; C is one of the model parameters ($C > 0$) to be determined; S represents the stress level; m is another model parameter, dependent on stress behaviour, also to be determined.
The power law model is simple when expressed or plotted in logarithmic form, where it becomes a straight line with the slope representing the value of parameter m, and the value of the intercept with the y-axis is a function of the constant S: (where the y-axis is the function)

$$\ln[L(S)] = -m\ln(S) - \ln(C). \quad (2)$$

The inverse power law is applicable to all distributions regularly used in reliability. The test acceleration factor is then:

$$A_{S\_IPL} = \frac{L(S_{Use})}{L(S_{Test})} = \frac{C^{-1}\times S_{Use}^{-m}}{C^{-1}\times S_{Test}^{-m}} = \left(\frac{S_{Test}}{S_{Use}}\right)^{m} \quad (3)$$

where: $A_{S\text{-}IPL}$ is the acceleration factor of stress by inverse power law; $L(S_{Use})$ is the life as a function of stress in actual use; $L(S_{Test})$ is the â function of stress applied in test.
Parameter C in the test acceleration cancels out, but the parameter m shall be determined for the item and the stress type. The parameter m in the inverse power relationship is a measure of the effect of the stress on the life. As the absolute value of m increases, so does the effect of the stress. Negative values of m indicate an increasing life with increasing stress. An absolute value of m approaching zero indicates small effect of the stress on the life, with no effect (constant life with stress) when $m = 0$.

2 The available test time: Choose the duration of the accelerated life test 360000 cycles.

3 The number of units subjected to accelerated life tests: suppose 20 specimens.

4 The distribution law of the number of cycles until failure used in accelerated life testing: choose Weibull distribution and give shape parameter.

5 The stress under normal use condition and maximum in accelerated condition: the twist angle in normal condition is 30° and the maximum angle is 50° (during the experiment only 40°).

6 The accelerated life test plan: choose three levels best compromise plan. This plan recommends three stress levels: a high stress level, which is the maximum allowable stress value that you specified during setup, a low stress level and a middle stress level.

The test plan was realized using the ALTA software, introducing the aforementioned parameters. The recommended test plan suggested stress levels to be used in the test and the recommended allocation of units to each stress level.

Table 1. Recommended test plan [1]

Stress level	Stress value [°]	Unit allocation [%]	Unit allocation [Qty]	Probability of failure [-]
Low stress level	40	47.3	9.46	0.211
Middle stress level	45	25	5	0.648
High stress level	50	27.7	5.54	0.99

The plan recommended 3 stress levels at 40°, 45°, 50° and number of units for each stress level. That was 9 units to test at twist angle 40°, 5 units at 45° and 6 units at 50°.

4 Experiment description

4.1 Characteristics of torsion bar

This chapter describes the practical implementation of the accelerated life test for the torsion bar of the combat track vehicle. The object of the test was a torsion bar of the track vehicle (thereafter called torsion bar). The section portrays the experimental test rig for imitating the operation of the torsion bar.

The suspension system of the track vehicle is designed to transfer the weight of the track vehicle through the track rollers and the track on the ground, to mitigate shocks and impacts acting on the track vehicle hull, and to quickly stop the hull oscillations. The quality of the suspension system to a large extent determines the average speeds of track vehicles on the ground, accuracy of fire on the move, crew performance, reliability and durability of the equipment of the track vehicle.

The track vehicle suspension system includes five suspension assemblies, each of which consists of a road arm, a torsion bar and a bumper. There are hydraulic absorbers on the first and fifth suspension assemblies. As a spring is used a torsion bar with a diameter of 52 mm, made of 45ChNMFA steel, processed by grinding and shot peening, followed by prestressing. Chemical composition and mechanical properties of the torsion bar material are shown in the Table 2.

Table 2. Chemical composition and mechanical properties of the torsion bar material [6]

C	S	P	Mn	Cr	W
0.42 - 0.50	≤ 0.025	≤ 0.025	0.50 - 0.80	0.80 - 1.10	≤ 0.20
V	**Ti**	**Si**	**Ni**	**Mo**	**Cu**
0.10 - 0.18	≤ 0.030	0.17 - 0.37	1.30 - 1.80	0.20 - 0.30	≤ 0.30
UTS MPa	**Yield MPa**	**Torsion limit MPa**	**Shear modulus GPa**	**Young's modulus GPa**	**Hardness HB**
1384	1080	800	80	205	269

4.2 Test rig construction

The test rig was built to test three torsion bars at three different twisting angles to determine the number of cycles to failure at every stress level. A complete test rig equipped with a complete measuring system has been designed and manufactured to measure the moment acting on torsion bar and the corresponding twist angle. The measured parameters are used to evaluate vehicle suspension characteristics. A lifetime of a torsion bar is predicted using the accelerated test in a fatigue test bench. Torsion bars involve repeated torsional loading. Torsional fatigue tests are performed on an axial-type machine. A typical torsional fatigue testing machine is shown in Fig. 2.

The test rig consists of a hydraulic power unit, hydraulic cylinder - actuator, and a controller. MTS SilentFlo™ Hydraulic Power Unit 505.30 delivers superior performance in servo hydraulic testing applications. The MTS Model 493.02 FlexTest SE Controller made by MTS is a fully digital Proportional, Integral, Derivative, Feedforward (PIDF) servo controller. It provides complete control of one servo hydraulic channel or station in an MTS test system.

The swing angle of the road arm is measured by the Gyro Enhanced Inclinometer FAS-G of the MicroStrain, Inc. The FAS-G not only has the ability to measure static angles, but also dynamic, fast angular movements. Through the use of the two accelerometers and one piezo-ceramic gyro coupled with the requisite digital filtering and embedded software tracking algorithms, FAS-G provides dynamic response while maintaining the DC (static) measurement accuracy. As a result, during rapid angular movements, both the static and the dynamic components of the movement can be measured. The accuracy of the sensor is ±1 degree.

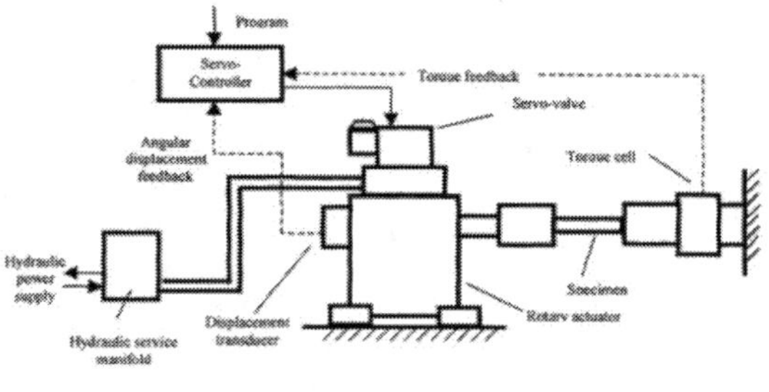

Fig. 2. Hydraulic torsional fatigue system [7]

A torsion bar was mounted on the test rig, one end was fixed, and the other end was connected with a swing arm, which rotated about the longitudinal axis of the torsion bar. The first torsion bar was twisted from 15° to 25° such as it operated in real life. Note that the static twist angle of the torsion bar is 14.8°. The second torsion bar was twisted from 15° to 30°. The range of the third torsion bar was from 15° to 40°. The first and the second torsion bars were swung with frequency 1.5 Hz while the third torsion bar was with frequency 1 Hz. Every torsion bar went through more than 360000 cycles and no failure was observed. The experimental results are obtained after measuring the applied torque on the torsion bar and the corresponding twist angle.

Table 3. Twist angle and torque of torsion bar

Twist angle, [°]	15	25	30	40	53
Twisting moment, [kNm]	7.7	13.0	15.5	20.6	-

5 Model of torsion bar suspension

While a track vehicle is travelling on road, it is subject to excitation from the road. Vertical motion of the road wheel causes the torsion bar to twist around its axis and is resisted by the bar's torsional resistance. The resistance of the torsion bar to twisting has the same effect as the coil spring used in conventional suspension systems.

From the model of torsion bar suspension as shown in Fig. 3, the vertical force exerted on the road wheel and transmitted through the axle arm to the torsion bar is determined by the formula [8]:

$$P_W = \frac{GJ}{L}\frac{\beta}{R\cos(\beta_0-\beta)} \quad (4)$$

where P_w - vertical force transmitted from road wheel to the hull; G - shear modulus of torsion bar material; J - polar second moment of torsion bar cross-section $J = \frac{\pi d^4}{32}$; d - diameter of torsion bar; β - twist angle of torsion bar; β_0 - setting angle of torsion bar; L - active length of torsion bar.

This force will cause a torsional moment acting on the torsion bar. The moment is correlated to the resulted twist angle β by the following relationship [8]:

$$T = P_W R\cos(\beta_0 - \beta) \quad (5)$$

therefore

$$T = \frac{GJ}{L}\beta = C\beta. \quad (6)$$

where T – twisting moment on torsion bar; R – radius of road wheel arm; C - torsional stiffness of the torsion bar.

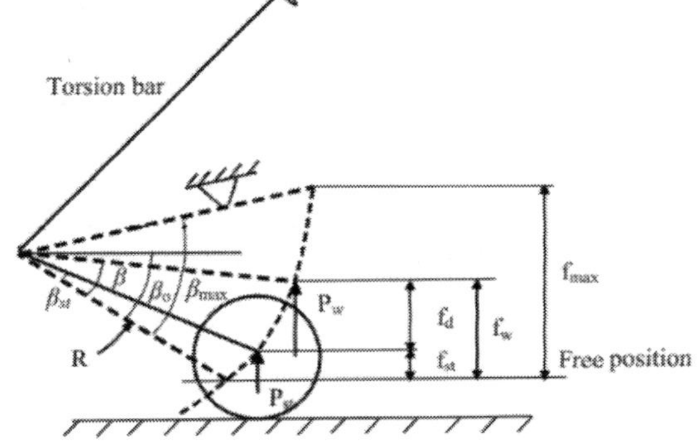

Fig. 3. Individual torsion bar suspension [8]

When a shaft is subjected to a torque or twisting, a shearing stress is produced in the shaft. The shear stress varies from zero in the axis to a maximum at the outside surface of the shaft. Shear stress β in the torsion bar is given by:

$$\tau = \frac{T}{J}r = \frac{Gr}{L}\beta \quad (7)$$

The torsional stiffness of a uniform rod is dependent on the cross-section diameter r, the length L, and the shear modulus G of elasticity of the rod. Hence, the torsion bar spring rate can be varied by appropriately changing any of the previous parameters [8].

In order to restrict the vertical travel of road wheel as well as safeguard the torsion bar, bumpers are welded on the track vehicle hull. Therefore, the twist angle of the torsion bar reaches the maximum value βmax when the swing arm touches the bumper. This angle is smaller than the one corresponding to the shear yield strength of the bar. On the track vehicle, this restrict angle accounts for 53°.

The static angle of the torsion bar βst is approximately 14.8°. The static and maximum twist angles of the torsion bar can be calculated from the schema of individual suspension or measured provided that the setting angle is β0.

The specifications of the individual torsion suspension are shown in the Table 4.

Table 4. Specifications of torsion bar suspension

Parameters	Values
Torsion bar diameter, [mm]	52
Torsion bar active length, [mm]	1960
Radius of road arm (swing arm), [mm]	250

6 Test rig simulation

In this article, the model of the test rig was built and tested using MSC Adams software. The virtual model allows simulating the torsion bar at the large twist angles that the test trig cannot. In this model, the torsion bar was represented by a 3D beam omprised of a single FE part of twelve nodes. One end is constrained with the ground and the other end is constrained with the swing arm.

Table 5. Values of twisting moment and shear stress at twist angle

	Twist angle		Torque, [kNm]	Shear stress, [MPa]	Equiv. stress, [MPa]
	angular [°]	[rad]			
Setting position	0	0	0	255	
Static angle	14.8	0.258	7.6	276	21
Static angle	25	0.436	12.9	466	211
Static angle	30	0.524	15.4	559	304

	Twist angle		Torque, [kNm]	Shear stress, [MPa]	Equiv. stress, [MPa]
	angular [°]	[rad]			
Static angle	40	0.698	20.6	745	490
Static angle	45	0.785	23.1	838	583
Static angle	50	0.873	25.7	931	676
Strict angle	53	0.934	27.5	996	741

The FE part is a wholly Adams-native modelling object with inertia properties and is accurate for very large deformation cases of beam-like structures. The FE part differs from the linear flexible body option within Adams Flex in two significant ways. Firstly, it has the ability to accurately represent large deformations which the linear modes approach cannot. Secondly, its modelling does not require an FEA produced file like the modal neutral file.

The FE Part also differs from the beam force element in that it possesses inertia properties [9].

For the test rig model and through the experiment, the following data were applied: torsion bar diameter - 52 mm, torsion bar active length - 1960 mm, shear modulus of torsion bar material - 80.23 kN/mm^2.

Table 6. Comparison of experimental and simulation results

Twist angle in [°]		15	25	30	40	53
Twisting moment kNm	Experiment [kNm]	7.7	13.0	15.5	20.6	-
	Simulation [kNm]	7.6	12.7	15.2	20.3	26.9

It is clear from the table that the results of physical tests and simulation have an insignificant difference.

Created model make possible to simulate the dynamic analysis, obtained the twist angle and the torque of torsion bar and compare to the real test results. The comparison is shown in the Table 6.

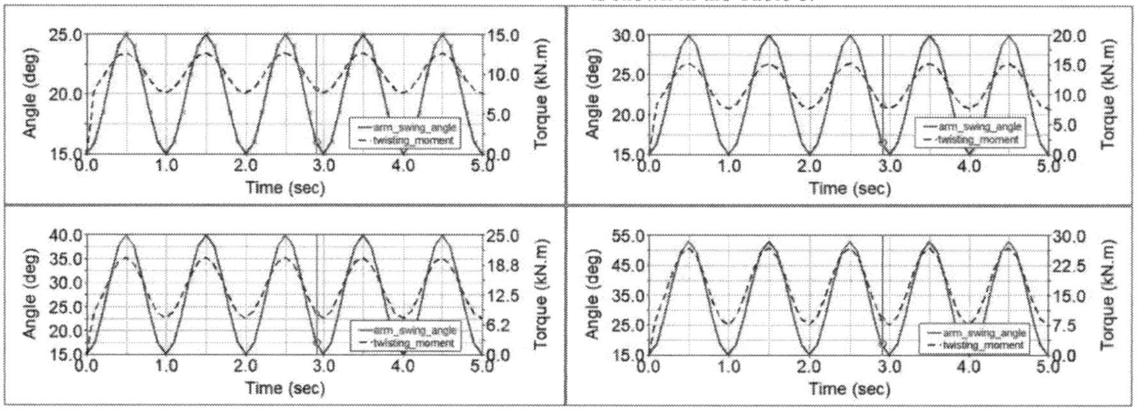

Fig. 4. Simulation results for twisting moment and swing angle of road arm

7 Conclusion

This article deals with the accelerated life test for a torsion bar of the track vehicle. Majority of mechanical damages of structural elements occur due to material fatigue during its normal operation. Fatigue is a major failure mechanism of mechanical parts. At the same time, this type of damage is considered to be the most dangerous one, the crack initiation in structural elements is difficult to predict, and it leads to damage of the elements.

The primary purpose of an ALT is to estimate the life distribution and quantities of interest in a use condition. This estimation involves extrapolation from higher stress levels by using an acceleration model and thus includes the model error and statistical uncertainty.

For the torsion bars of tracked combat vehicles with stringent functionality specifications, the reliability is very high. Therefore, their life testing under nominal conditions requires much time and resources. In response to this problem, ALT finds an application to obtain timely information on the reliability of the products.

References

[1] IEC 62506: Methods for Product Accelerated Testing. (2003)

[2] B. Bertsche, Reliability in Automotive and Mechanical Engineering - Determination of Component and System Reliability. (Springer, 2008)

[3] W. Q. Meeker, L. A. Escobar, Statistical Methods for Reliability Data. (Wiley, 1998)

[4] W. B. Nelson, Accelerated Testing - Statistical Models, Test Plans, and Data Analysis. (Wiley, 2004)

[5] G. Yang, Life Cycle Reliability Engineering. (Wiley, 2007)

[6] B. Leitner, L. Figuli, *Fatigue life prediction of mechanical structures under stochastic loading*. **157**. (EDP Sciences, Les Ulis, 2018)

[7] ASM International. Handbook Vol. 8: Mechanical Testing. The Materials Information Society. (1996)

[8] L. V. Sergeev, *Theory of tank*. Academy BTV Press, Moscow, 1973.

[9] Mechanical dynamics. Building Models in ADAMS/View. (Ann Arbor, Michigan, 2000)

Study on Sealing Performance of Oil Seal with Micro-pores Textured in Rotary Shaft Surface

CHENG Xiang-ping*, *ZHANG* You-liang, *KANG* Lin-ping

Institute of Applied Physics, Jiangxi Academy of Sciences, Nanchang 330029, China)

Abstract. In order to improve the reliability and the long life cycle of the lip seals, comprehensive consideration of the liquid film cavitation and JFO mass conservation boundary conditions, the geometry model of the oil seal with micro-diamonds textured on the shaft surface is given, and its mathematical model is built, the relevant numerical calculation method is used, eventually that the film pressure distribution and the micro diamond pores structure parameters effect on the seal performance are obtained under the different operating conditions. The results showed that the dynamic pressure effect caused by the diamond pores can make the oil film pressure field between the axial surface and the lip produced regular axial and radial wave change. At the same time, the change of working conditions strengthen or weaken the change law of the oil film pressure field, so which make the reliability of the liquid pressure, lubricating property and pump suction effect also change accordingly. The size, shape and pores direction of the micro-diamonds texture have great influence on the oil seal performance and lubrication characteristics, which can help to reduce the leakage rate, control the pump suction direction, stable the liquid pressure and reduce the friction. In order to improve the stability of the liquid film and pump suction effect, reduce the leakage rate, friction and wear, the axial surface micro-diamond pores texture of which the cone width ratio is 0.4-0.6, the pores depth is 1.5-4.5μm and the tilt angle is 40°-50° shall be selected for the oil seals.

In recent years, with the continuous improvement of many requirements for the lubrication characteristics of the key friction and wear parts of the rotating equipment, the micro-textured technology of the contact surfaces has developed rapidly with its excellent lubrication performance. At present, with the growing expansion of high precision, cutting-edge industry, the requirements of which the environmental protection and energy saving, seal reliability and long life cycle of the types of lip seals are also more and more high, so need to design and develop long use period and better performance of the new oil seals, which become a problem to be solved, lip seals with axial surface micro forming technology have been put on the agenda, and incomparable better performances than traditional oil seals are found, the rotate equipments in agricultural machinery, aerospace, marine, petroleum and petrochemical industries have been gradually put into use, and the effect is excellent. Domestic and foreign scholars[1-7] had also carried out a large number of theoretical and experimental studies to reveal the micro-woven structure of the axial surface and improve the friction lubrication mechanism between contact surfaces, effectively improving the lubrication performance. Moreover, it was found that the shape, size, pores depth, pores orientation and area ratio of axial surface micro-texture[8-14] had greatly influenced on the lubrication performance of the oil seal. In addition, the mechanism of the surface micro-modeling to which improve the lubrication state and the sealing performance is very complex, which is not only dependent on its shape and structure, but also has a great relationship with working conditions, and its internal mechanism needs to be further revealed. In view of this, aimed at the different coaxial surfaces microstructure textures as the research object, which mass conservation Reynolds governing equation is built between sealing surfaces in this paper, and the distribution of film pressure was analyzed, the influence of structural parameters and operating parameters of the micro-model textures on the sealing performance of micro-pores shaft surface was studied, its lubrication mechanism was delved into.

1 Model

1.1 Structure model

Fig.1 shows the assembly structure and local profile of typical axial micro-pores texture oil seal, and the micro-pores can be triangles, squares, rectangles, ellipses, diamonds and circles, etc. Along the axial and circumferential directions, the suitable micro-pores are uniform etched in the axis surfaces. However, there are only two rows of micro-pores in the axial direction

* Corresponding author: 654268480@qq.com

 https://doi.org/10.1051/matecconf/202032703002

between the lip and axial surface, L is the axial length in sealing area, h_0 is the balance film thickness, h1 is the pores depth, a and b are respectively the effective half side length in the calculation formula for the area of micro-pores, the side length of circulars and squares is $a=b=r$, the shape factor of all micro-pores is $\gamma=b/a$ except the circles and squares. The angle between the semi-long axis and the tangent direction of the rotating axis through the pore center is α_1.

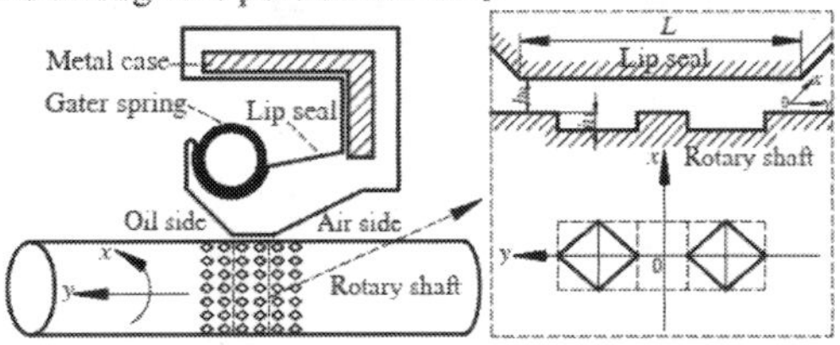

(a) micro-pores distribution of shaft surface
(b) Schematic diagram of micro-pores in a cycle

Fig.1 Schematic of the lip seal structure and local profile

1.2 Mathematical model

It is assumed that the sealing medium is Newtonian fluid, and its viscosity and pressure remain constant along the direction of liquid film thickness. In consideration of liquid film cavitation, JFO mass conservation Reynolds control equation is adopted:

$$\frac{1}{r}\frac{\partial}{\partial\phi}\left(\frac{h^3}{12\mu}\frac{\partial p}{\partial\phi}\right)+\frac{\partial}{\partial r}\left(\frac{rh^3}{12\mu}\frac{\partial p}{\partial r}\right)=\frac{rU}{2}\frac{\partial(\theta h)}{\partial\phi} \quad (1)$$

Type in the

$\theta=\rho/\rho_l$(ρ-the medium density, ρ_l-the medium density in non-cavitation region), in the non-cavitation region: $\theta=1$, $p>p_{cav}$, in the cavitation region: $0<\theta<1$, $p=p_{cav}$. h is the liquid film thickness; U is the rotating shaft speed; p and p_{cav} are respectively oil pressure and cavitation pressure.

Assuming that the lip part and the axial surface are smooth contact surfaces, the liquid film thickness equation of the sealing area is as follows:

$$h=\begin{cases} h_0 & \text{pores outside} \\ h_0+h_1 & \text{pores inside} \end{cases} \quad (2)$$

Since the micro-pores are arranged periodically on the rotating shaft surface, the single period is selected as the research object in Fig.1.

The mandatory boundary condition:

$$\begin{cases} P\left(-1\frac{1}{2},\ Y\right)=P_o \\ P\left(1\frac{1}{2},\ Y\right)=P_a=P_i \\ \theta\left(-1\frac{1}{2},\ Y\right)=\theta\left(1\frac{1}{2},\ Y\right) \end{cases} \quad (3)$$

Periodical boundary conditions:

$$\begin{cases} P\left(X,-\frac{1}{2}\right)=P\left(X,\frac{1}{2}\right) \\ \theta\left(X,-\frac{1}{2}\right)=\theta\left(X,\frac{1}{2}\right) \end{cases} \quad (4)$$

JFO cavitation conditions:

$$\begin{cases} p>p_c,\theta=1 \\ p<p_c,0<\theta<1 \end{cases} \quad (5)$$

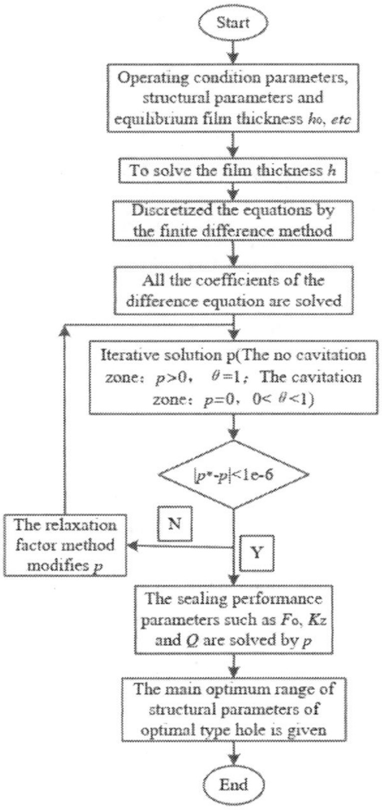

Fig.2 Flow chart for computation

The calculation process is shown in Fig.2, the process is as follows:

①Input the base film thickness, geometric and operating parameters;

②Combining JFO mass conservation boundary conditions, the finite difference method is used to transform equation (1) into a difference equation by discrete decomposition, and the initial pressure p_o is used to solve all the coefficients of the difference equation.

③In the process of iteratively calculating pressure p, the relaxation iteration method is used to correctly calculate p. After repeated iterations, until the pressure convergence criterion conforms to the given value, the iteration is stopped and the next step is continued.

④The finally solved pressure is used to calculate parameters such as the liquid film stiffness, opening force and leakage rate, etc. The specific algorithm is shown in the literature[15].

2 Calculation results and discussion

(1) Structural parameters: L=100μm, h_0=2.0μm, h_1=4 μm, pores number in per cycle n_s=2, γ=0.5, a=15μm, α_1=0.

(2) Operating parameters: the oil side pressure p_o=0.12 MPa, air side pressure p_a=p_i=0.1MPa, cavitation pressure p_c=0, shaft rotary speed U=200r/min, viscosity of oil μ=0.01Pa.s.

When calculating and analyzing the influence of the structure and operation parameters on the sealing performance, the other geometric and the operating parameters remain unchanged, except as specified.

2.1 Influence of the operating parameters on the sealing performance

2.1.1 Impact of po

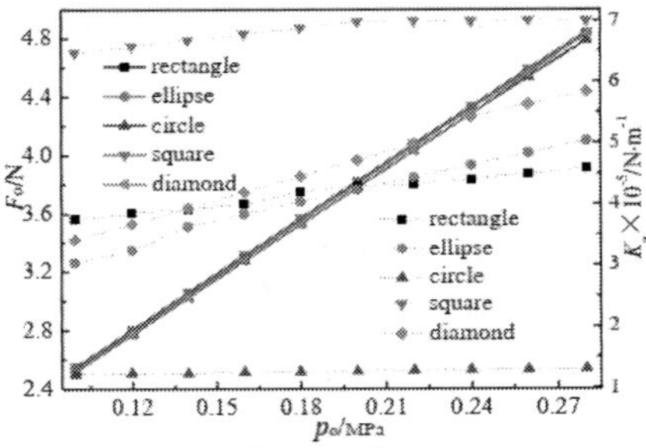

(a) F_o、K_z&p_o

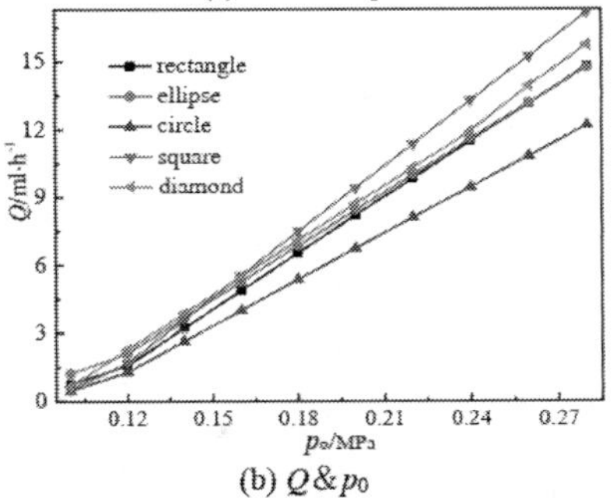

(b) Q&p_o

Fig.3 Oil film pressure diagram under the change of p_o

It is shown that the influence on the sealing performance of each pore texture in oil seal zone in the Fig.3 when p_o changes, the solid curve in the figure corresponds to the vertical axis on the left, while the virtual curve corresponds to the vertical axis on the right (Fig.3(a)), the x-axis is shared. With the continuous increase of the pressure, it is shown that the K_z and F_o of the other pores texture lip seals show an increasing tendency, except that the liquid film stiffness of the circular texture pores tends to be horizontal linear, while the leakage rate of all pores increases constantly with the increase of p_o, that is, the stability of the liquid film (except the circular pores) and the bearing capacity become larger and larger, and the leakage ratio is becoming more and more big with the increase of p_o.This indicates that the static pressure effect is increasing with the pressure difference increasing, and the leakage rate of the all pores also increases with the increase of the oil side pressure, while the dynamic pressure effect of axial surface micro-pores remains unchanged under the constant speed condition, under the combined action of dynamic, static pressure effect and shear effect, the pump drain effect of axial surface pores weakens, resulting in the continuous increase of the leakage rate. It can also be seen that the open force of each pore is almost same under the same pressure. Q is sorted as $Q_{(square)}>Q_{(diamond)}>Q_{(rectangle)}>Q_{(ellipse)}>Q_{(circle)}$. When the oil side pressure is less than 0.14MPa, K_z is sorted as $K_{z(square)}>K_{z(rectangle)}>K_{z(diamond)}>K_{z(ellipse)}>K_{z(circle)}$. When the scope of p_o is 0.14MPa<p_o<0.21MPa, K_z is sorted as $K_{z(square)}>K_{z(diamond)}>K_{z(rectangle)}>K_{z(ellipse)}>K_{z(circle)}$. When the p_o is greater than 0.21MPa, K_z is sorted as $K_{z(square)}>K_{z(diamond)}>K_{z(ellipse)}>K_{z(rectangle)}>K_{z(circle)}$.

2.1.2 Influence of U

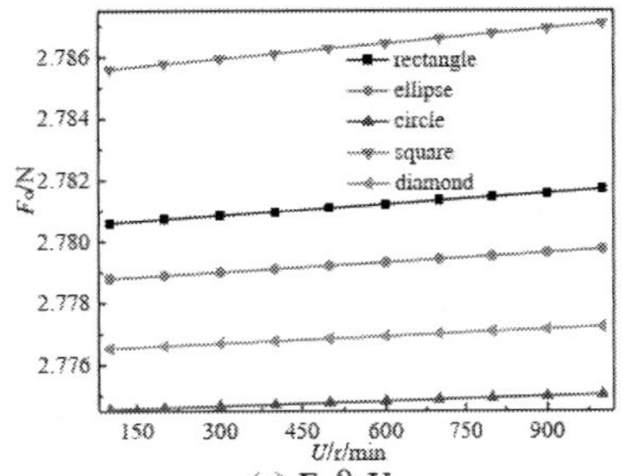

(a) F_o&U

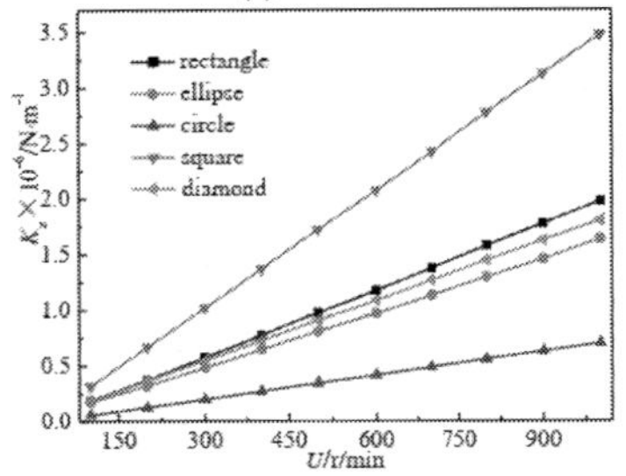

(b) K_z&U

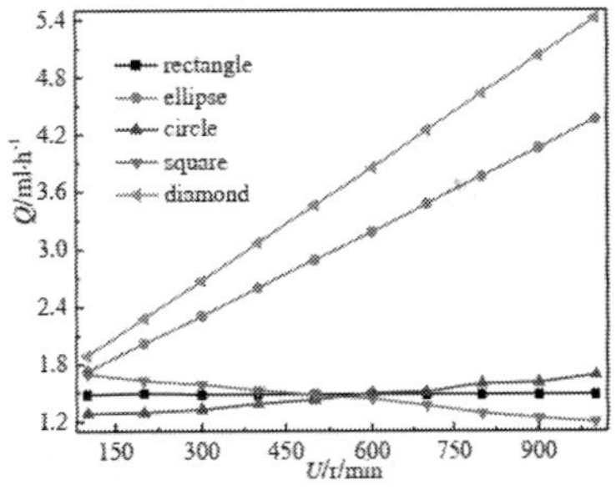

(c) Q&U

Fig.4 Effect of U on sealing performance parameters

It is shown that the influence law of rotary speed on the sealing performance of each pore texture in oil seal zone in fig.4. As the chart shows, K_z and F_o of each pore texture show a trend of linear increase with the increase of the rotary speed, and the change range is larger, except for the square, the Q of other pores increases with the increase of rotating speed. The reason is that the hydrodynamic pressure effect and pump drain effect in the sealing area will increase significantly with the increase of the rotary speed, therefore, the bearing capacity and stability of the liquid film are constantly strengthened. In addition to the square micro-pores, the Q of the other micro-pores also increases, however, the increase of pump drain effect caused by square micro-pores with the increase of velocity is stronger than that caused by centrifugal force. Therefore, the media which about to be released from the sealing area will be pumped back to the oil side in large quantities, so Q is going to decrease as speed goes up. It can also be seen that F_o and K_z are sorted as follows: $F_{o(square)}>F_{o(rectangle)}>F_{o\,(ellipse)}>F_{o(diamond)}>\ F_{o(circle)}$ and $K_{z\,(square)}>K_{z(rectangle)}>K_{z(diamond)}>K_{z(ellipse)}>K_{z(circle)}$. When Q is less than 570r/min, $Q_{(diamond)}>Q_{(ellipse)}>Q_{(circle)}>Q_{(rectangle)}>Q_{(square)}$. In a word, considering K_z and Q comprehensively, square and triangle micro-pores shaft oil seals have better sealing and lubrication effect under the same working condition.

2.1.3 The influence of h_0

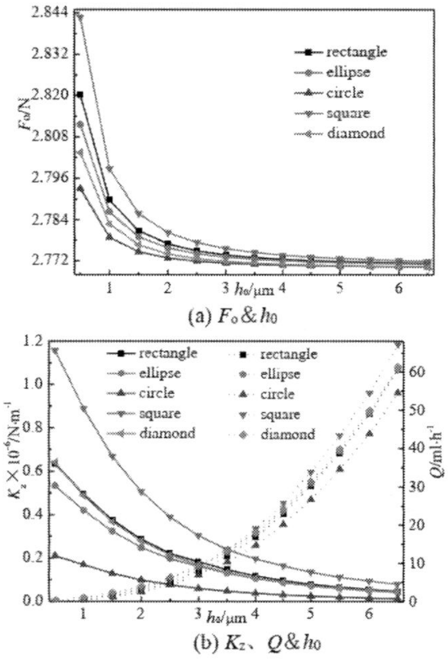

Fig.5 Effect of h_0 on sealing performance parameters

The fig.5 shows the influence law of h_0 change on the sealing performance parameters in the sealing area. As can be seen from the figure, with the increase of h_0, the K_z and F_o of each axis surface micro-pores oil seal show a decreasing trend, while Q gradually increases. When h_0 is smaller, K_z and F_o have a large change range, while Q has a small change range. With the increase of h_0, the change amplitude of K_z and F_o decreases, and the change amplitude of Q becomes larger. It is indicated that h_0 has a great influence on the friction lubrication characteristics and sealing performance of the oil seal with micro-pores in the axial surface. Due to the influence of mixed friction, the lubrication and sealing performance of the liquid film will be better when h_0 is relatively small. When h_0 is larger, it gradually enters the whole liquid film lubrication state, which results in a gentle decline in the lubrication characteristics and the stability of liquid film, and Q increases. When the h_0 is the same, it can also be seen that the K_z and Q of each oil seal with micro-pores in the axis surface decrease successively as: square, diamond, ellipse, triangle and circle, and the F_o of each oil seal with micro-pores is $F_{o(square)}>F_{o(rectangle)}>F_{o(ellipse)}>F_{o(diamond)}>F_{o(circle)}$. In conclusion, it is the best choice for better sealing performance when h_0 is 1.5-4μm.

2.2 Influence of structure parameters on sealing performance

2.2.1 Influence of γ

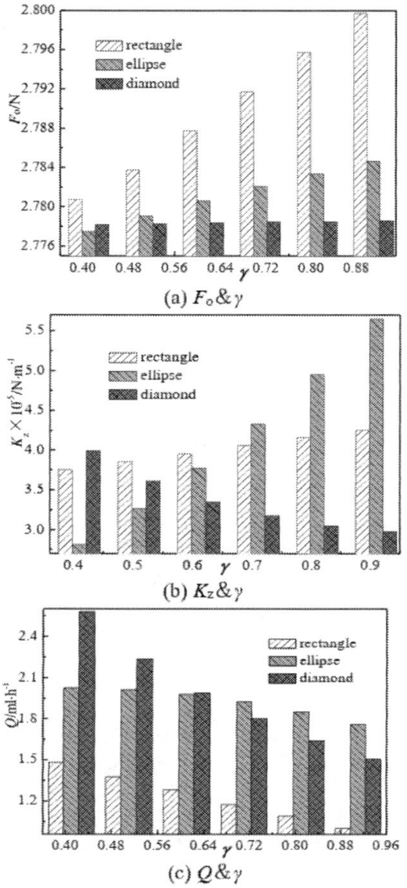

Fig.6 Effect of γ on sealing performance parameters

It is shown that the influence law of γ on the sealing performance of the oil seals in fig.6. With the increase of γ, the F_o of the oil seals with each micro-pore in shaft surface increases gradually, and the Q decreases gradually. $K_{z(rectangle)}$ and $K_{z(ellipse)}$ increase gradually, and the $K_{z(diamond)}$ decrease gradually. When γ is the same, it can also see that the F_o and Q are ordered as follows: $F_{o(rectangle)}>F_{o(ellipse)}>F_{o(diamond)}$ and $Q_{(rectangle)}<Q_{(ellipse)}<Q_{(diamond)}$. When γ is greater than 0.6, $K_{z(ellipse)}>K_{z(rectangle)}>K_{z(diamond)}$. To sum up, for oil seals, if it is required that the liquid film stability and sealing in the sealing area are more reliable, the texture oil seals of the diamond micro-pores when γ is 0.5-0.6 and triangle micro-pores axial when γ is 0.8-1.0 should be selected.

2.2.2 Influence of h1

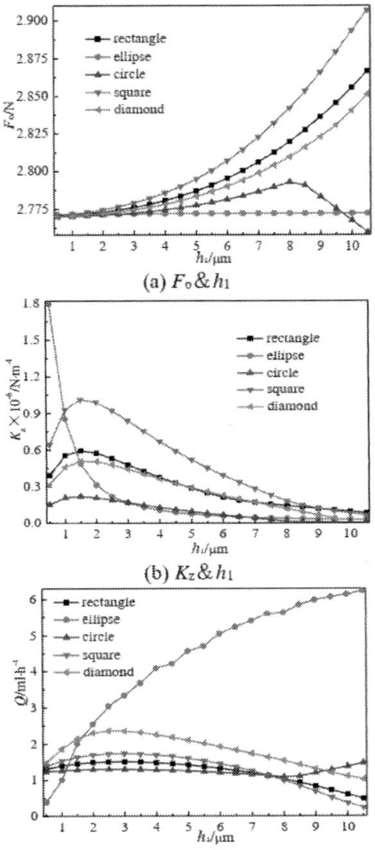

(a) F_o&h_1

(b) K_z&h_1

(c) Q&h_1

Fig.7 Effect of h_1 on sealing performance parameters

It is shown that the effect of h_1 on lip seal performance in the fig.7. Results show that with the increase of h_1, the open power of other micro-pores shows an increasing trend, except that the open power of circular micro-pores increased first and then decreased. As for the liquid film stiffness and leakage rate, with the increase of pores depth, except that the liquid film stiffness of the elliptic pores oil seal gradually decreases and the leakage rate increases, the liquid film stiffness and leakage rate of axial surface oil seals of other micro-pores all show a change trend of increasing first and then decreasing. And the liquid film stiffness has a maximum value at h_1=1.5-2.5μm, while the leakage rate has a maximum value at h_1=2-4μm. With the increase of pores depth, It is suggested that the dynamic pressure effect of the liquid film increases significantly, because of the strong cavitation, shear and extrusion effects of the liquid film in the initial stage, when the pores depth increases to a certain extent, the fluid hydrostatic effect in the micro-pores increases and the squeezing effect decreases, resulting in the reduction of cavitation effect, as a result, the liquid film stiffness and leakage rate increase first and then decrease. Considering the stability of the liquid film in the sealing area and the sealing reliability, the axial surface micro-pores texture interval should be selected when the pores depth is 1.5-4.5μm.When the pores depth is the same, it can also be seen that the open power and liquid film stiffness ordering of other micro-pores except for elliptical pores are basically as follows:$F_{o(square)}>F_{o(rectangle)}>F_{o(diamond)}>F_{o(circle)}$ and $K_{z\ (square)}>K_{z\ (rectangle)}>K_{z\ (diamond)}>K_{z\ (circle)}$. The order of leakage rate is basically $Q_{(ellipse)}>Q_{(diamond)}>Q_{(square)}>Q_{(rectangle)}>Q_{(circle)}$. In a word, combined with the sealing performance of each oil seal with micro-pores, the best effect is square, followed by triangle, diamond, circle and ellipse.

3 Conclusion

(1) the micro-pores texture in the axial surface can make the liquid film in the sealing area produce significant dynamic pressure effect, which can make the contact surface of the lip seal quickly detach from the axial surface, and enter into the lubrication state of the whole film, so as to avoid contact friction, thus reducing the strict requirements on the roughness and flatness of the axial surface. At the same time, part of the axial surface micro-pores can produce pump drain on oil, the leaking liquid would be back to the oil side, so the sealing reliability can be improved.

(2) With the increase of p_o, the static pressure effect will be increased which make the K_z, Q and F_o increase, and even cause oil seal failure due to leakage. With the increase of U, the bearing capacity and stability of the liquid film were enhanced. When h_0 increases, the Q, K_z and F_o of oil seal with diamond micro-pores in the shaft surface reduce, and the sealing performance is the best when h_0 is 1.5-4μm.

(3) Considering the stability of the liquid film and sealing reliability in the sealing area, axial texture oil seals with micro-diamond pores of which h_1 is 1.5-4.5μm and γ is 0.5~0.6, or triangular micro-pores of which γ is 0.8~1.0 should be chosen as the best seal under study condition. The best sealing performance of each micro-pore is square, followed by triangle, diamond, circle and ellipse.

Fund: the National Regional Science foundation of China (No. 51865019); Natural Science Foundation of Jiangxi Province (No. 20201BBE51015)

References

1. HORVE L A. The correlation of rotary shaft radial lip seal service reliability and pump ability to wear track roughness and micro-asperity formation[R]. SAE paper, 910530, 1991.

2. BAART P, LUGT P M, PRAKASH B. Review of the lubrication, sealing, and pumping mechanisms in oil and grease-lubricated radial lip seals [J]. J. engineering tribology, 2009, 223(3): 347-358.
3. JIANG Hua-sheng, MENG Xiang-kai, SHEN Ming-xue, PENG Xu-dong. Hydrodynamic lubrication performance of lip seal with inclined micro-pores manufactured on rotary shaft surface [J]. CIESC journal, 2015, 66(2): 678-686.
4. GABELLI A, POLL G. Formation of lubricant film in rotary sealing contacts (I): Lubricant film modeling [J]. ASME J. tribology, 1992, 114: 280-287.
5. PAIGE J, STEPHENS L S. Surface characterization and experimental design for testing of radial lip seal [J]. Tribology trans., 2004, 47(3): 341-355.
6. HADINATA P C, STEPHENS L S. Soft elasto-hydrodynamic analysis of radial lip seals with deterministic micro-asperities on the shaft [J]. ASME J. tribology, 2007, 129(4): 851-59.
7. WARREN K H, STEPHENS L S. Effect of shaft micro-cavity patterns for flow and friction control on radial lip seal performance a feasibility study [J]. Tribology trans., 2009, 52(6): 731-743.
8. KANAKASABAI V, WARREN K H, STEPHENS L S. Surface analysis of the elastomers in lip seals run against shafts manufactured with micro-cavity patterns [J]. J. engineering tribology, 2010, 224(8): 723-736.
9. JIANG Hua-sheng, PENG Xu-dong, MENG Xiang-kai, SHEN Ming-xue. Numerical analysis of reverse pumping rate of rotary lip seal by finite element method[J]. CIESC journal, 2014, 48(8):1194-1199.
10. ZHOU Qiong. Research on the Lubrication Performance of Lip Seal and the Influences on the rotor dynamics performance[D]. East china university of technology, 2013.
11. GUO Fei, JIA Xiao-hong, HUANG Le, WANG Yu-ming. A mixed lubrication theoretical model and experimental verification of rotary lip seals[J]. Journal of mechanical engineering, 2014, 50(3):137-144.
12. SALANT R F. Soft elasto-hydrodynamic analysis of rotary lip seals [J]. J. mechanical engineering science, 2010, 224(12): 2637-2647.
13. Li W,STEPHENS L S,WENK J F. Experimental benchmarking of the numerical model of a radial lip seal with a surface textured shaft[J].Tribology transactions, 2013, 56(1): 75-87.
14. MENG Xiang-kai, BAI Shao-xian, PENG Xu-dong. Lubrication film flow control by oriented dimples for liquid lubricated mechanical seals [J]. Tribology international, 2014(77):132-141.
15. Wen Shi-zhu, Huang Ping. Principles of tribology[M]. Beijing: Tsinghua University Press, 2008: 54. (in Chinese)

MATEC Web of Conferences 327, 03003 (2020) https://doi.org/10.1051/matecconf/202032703003
ICMIE 2020

Kurtosis Based Empirical Mode Decomposition for Rolling Bearing Fault Detection

Hui Li* and *Xuhan* Liu

School of Mechanical Engineering, Tianjin University of Technology and Education, Tianjin, China

Abstract. A bearing fault diagnosis approach based on spectral kurtosis and empirical mode decomposition (EMD) is proposed. EMD is a signal decomposition technique, which can adaptively separate a number of intrinsic mode functions (IMFs) from the vibration signal according to the architectural characteristics of the data. The spectral kurtosis parameter takes as signal impulsive indicator. Firstly, EMD is utilized to process the sampling vibration signal. And then spectral kurtosis is calculated to select the optimal intrinsic mode functions, so as to suppress the noise and highlight the transient impact feature. Finally, the envelope spectrum is computed and the fault characteristic is recognized. The experimental results show that the proposed approach can identify bearing defects effectively and provide a reliable method for gearbox fault monitoring and diagnosis.

1 Introduction

Bearings are important mechanical components in various mechanical equipments, such as automobile, metallurgical equipment, etc. The unexpected failure of bearings will affect the normal operation of the equipment, and even cause significant economic losses. Therefore, bearing fault detection and diagnosis has been a hot topic for scholars. Among many methods of bearing fault feature extraction, the approach based on vibration signal analysis technique is widely utilized because of its high reliability. For example, fast Fourier transform (FFT), power spectrum estimation, envelope spectrum and cepstrum analysis technique have achieved good results in bearing fault detection and diagnosis [1,2]. However, All of these signal processing technique are all in view of the hypothesis of linearity and stationarity of vibration signals. But the vibration signal of rolling bearing defect has the characteristics of non-linear and non-stationary, and the vibration signal of bearing fault has the characteristics of transient impact in essence. Therefore, the rolling bearing system defect recognition and diagnosis need to study new vibration analysis technology.

This paper put forwards an approach of bearing fault monitoring and recognition based on EMD and spectrum kurtosis based envelope spectrum. EMD is a new non-stationary, nonlinear adaptive signal decomposition method proposed by Huang [3]. EMD can adaptively separate multi component signals into a number of intrinsic mode functions (IMF). Then the spectrum kurtosis is utilized to select the optimal parameters of band-pass filter, which can effectively suppress the noise in the signal and highlight the transient impact characteristics of the signal. Finally, the envelope spectrum of the selected band-pass signal is analyzed, and the bearing fault is identified and diagnosed according to the spectrum characteristics of the envelope spectrum. The envelope spectrum method based on spectral kurtosis can effectively extract the feature frequencies related to bearing defects. The basic principle of the approach is presented briefly, and the method is applied to bearing defect monitoring and recognition. Experimental results show that this approach can effectively monitor and identify rolling bearing faults.

2 The theory of kurtosis based EMD

In 1998, Huang et al [3,4] defined the IMFs and put forward the EMD technique. EMD is an effective technique for nonlinear and non-stationary signal processing.

2.1 Empirical mode decomposition

EMD method is put forward based on the hypothesis of simple morphological diversity, that is, any vibration signal is composed of function superposition with various of simple natural vibration modes. The innate character of this technique is to use the characteristic time scale of the vibration signal to identify the intrinsic vibration mode of the signal, and then the signal is adaptively decomposed into the sum of several IMFs. The characteristic time proportion can be defined by the time difference between two consecutive extremes.

In order to pick up IMF adaptively from a vibration signal, the signal sifting process is given as bellow. Firstly, all local extrema of vibration signal, including local maximum and local minimum, are determined, and

* Corresponding author: huili68@163.com

then all local maxima are fitted together by cubic spline as upper envelope. Then, the local minima are fitted together to generate the lower envelope of the vibration signal. The upper and lower envelope lines of vibration signals shall contain all points between them. Their average value is defined as $f_1(t)$, the difference between the data and $f_1(t)$ is defined as $d_1(t)$, i.e.:

$$x(t) - f_1(t) = d_1(t) \tag{1}$$

In an ideal situation, $d_1(t)$ should be an intrinsic mode function, because the methods and steps of sifting $d_1(t)$ represented above should force the results to meet all conditions of the IMF by constructing $d_1(t)$. To check whether h is an IMF, $d_1(t)$ only needs to meet the following conditions: (i) $d_1(t)$ should not be affected by riding waves, that is, the first component should not be under-shoot or over-shoot displayed on the data, and local extremum should be generated without zero crossing. (ii) The displayed upper and lower envelope lines are symmetrical with regard to zero. (iii) Clearly, the number of zero crossing points and extreme points should be the same in these two functions.

Such a data filtering process may be repeated many times to ensure that the extracted signal is an IMF. In the subsequent sifting, $d_1(t)$ is treated as the data to be processed, as follows:

$$d_1(t) - f_{11}(t) = d_{11}(t) \tag{2}$$

Where $f_{11}(t)$ is the average value of the upper envelope and the lower envelope of $d_1(t)$.

The above mentioned process should be redone k times and $d_{1k}(t)$ is given as follow

$$d_{1(k-1)}(t) - f_{1k}(t) = d_{1k}(t) \tag{3}$$

After each filtering step, it must be checked whether the number of zero crossing points and the number of extreme values are exactly the same.

The final shifting result is the first IMF, and then it is appointed as $c_1(t) = d_{1k}(t)$. The first IMF element from the data should embody the highest vibration frequencies searched in the original signal.

Then subtract the first IMF from the original signal, and the difference, is called a residue $r_1(t)$ which is formulated as follow:

$$x(t) - c_1(t) = r_1(t) \tag{4}$$

The residue $r_1(t)$ is taken as if it was the original data and we apply to it again the sifting process. The data shifting process of finding more intrinsic modes $c_i(t)$ will be repeated many times to find all the intrinsic mode functions in the signal until the last one is found. The final residue will be a monotonic function or a constant. In general, The residue will be the general trend of data.

$$x(t) = \sum_{i=1}^{n} c_i(t) + r_n(t) \tag{5}$$

After the above steps, the original data is effectively separated into n number of intrinsic mode functions and a residual, which can be an average trend or a constant [5].

2.2 Spectral kurtosis

Spectral kurtosis is the fourth-order statistic of the signal, which is mainly used to calculate the spectral kurtosis in the frequency spectrum map, so as to highlight the transient impulsive characteristics of the primitive vibration data. This approach conquers the shortcoming that the Fourier transform can't monitor and identify the transient impact feature of original sampled vibration data.

Spectral kurtosis represents the peak of probability density function, which can be formulated as follow

$$K_x(f) = \frac{S_{4x}(f)}{S_{2x}^2(f)} - 2 \tag{6}$$

Where $S_{4x}(f)$ is the fourth order cumulant of the spectral density of the original vibration data $x(t)$ and $S_{2x}(f)$ is the second order cumulant.

When the sampled vibration data $x(t)$ contains noise, the spectral kurtosis is given as follows:

$$K_s(f) = \frac{K_x(f)}{|1+\rho(f)|^2} \tag{7}$$

Where $K_x(f)$ is the spectral kurtosis of original vibration data $x(t)$ without noise and $\rho(f)$ is the count backwards of signal to noise ratio.

The magnitude of the kurtosis reflects the severity of the transient shock phenomenon in the original vibration data. It is related to the resonance frequency band of the fault vibration signal of the selected bearing. The premise of accurately extracting and identifying bearing fault features is to find the frequency band center and its bandwidth where the peak value of spectrum lies.

3 Bearing damage detection based on kurtosis based EMD

The experimental ball bearing is only utilized to research one kind of localized surface defect: bearing outer or inner raceway defect. The tested rolling bearing is used discharge device to machine a slot on the inner or outer race. Then the localized fault is a seed formed on the outer or inner raceway. The defect was 1 mm deep and 1.5 mm wide. The gearbox is driven by three-phase alternating current motor. The input spindle speed is 1500 r/min, that is to say, f_r is 25Hz. The bearing model is deep groove ball bearing 6208. The number of balls is 10 and the contact angle $\alpha = 0°$. The bearing pitch diameter D is 97.5mm and the ball diameter d is $55/3$ mm. The rolling bearing defect characteristic frequencies of inner race is 148.5Hz [6].

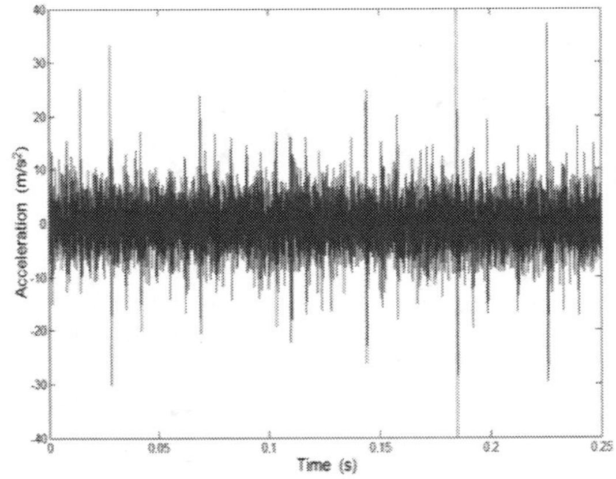

Fig. 1. Vibration data with inner defect.

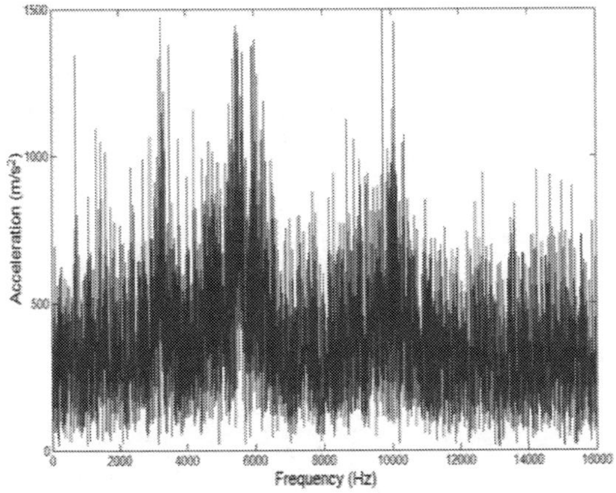

Fig.2. FFT with inner defect.

The sampled bearing inner race defect vibration data is displayed in Fig.1 and the FFT is showed Fig.2. In Fig. 2, there is no significant spectral peak near the rolling bearing defect characteristic frequency f_{inner} and its higher harmonics components. Therefore, the classical Fourier analysis can not accurately extract the transient impact characteristics of bearing fault in noise environment, so it is difficult to effectively diagnose bearing localized fault.

The EMD algorithm is utilized for the vibration data of Fig.1. Fig.3 gives the empirical mode decomposition of vibration data with inner race defect, which has eleven intrinsic mode functions. There are eleven intrinsic mode functions: c_1 - c_{13}, which selected from empirical mode decomposition. The IMFs represent the intrinsic modes of various frequency components actuated by bearing inner race fault, c_{13} is the residue, respectively. Among them, c_1 has the highest frequency of the vibration data, c_2 represents the next higher frequency component, and others. These empirical modal functions clearly describe the transient vibration feature caused by bearing inner race fault. It can be easily proved from Fig. 3 that EMD is very effective in the decomposition of vibration signals.

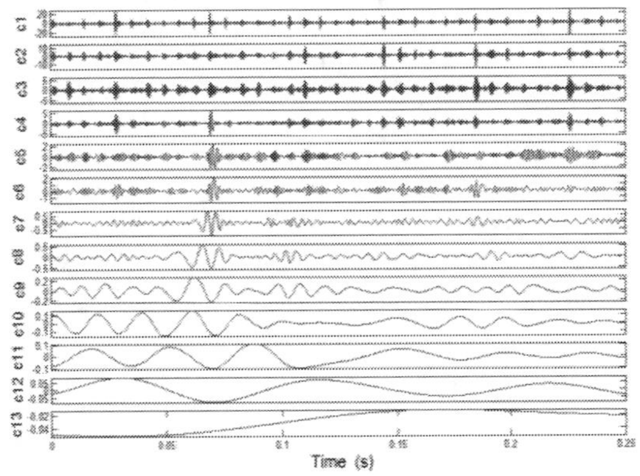

Fig. 3. IMFs of the data with inner defect.

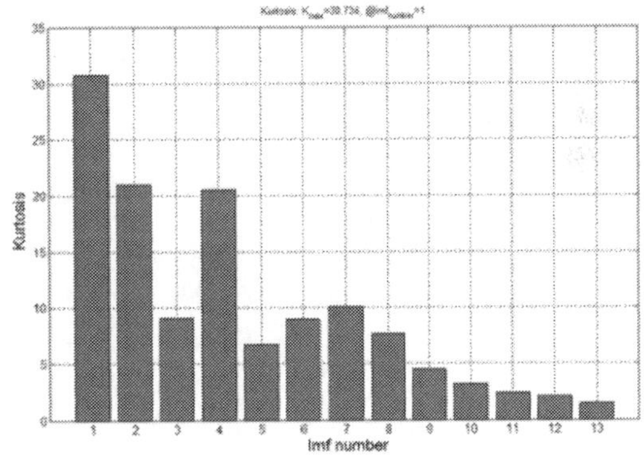

Fig. 4. Spectral kurtosis.

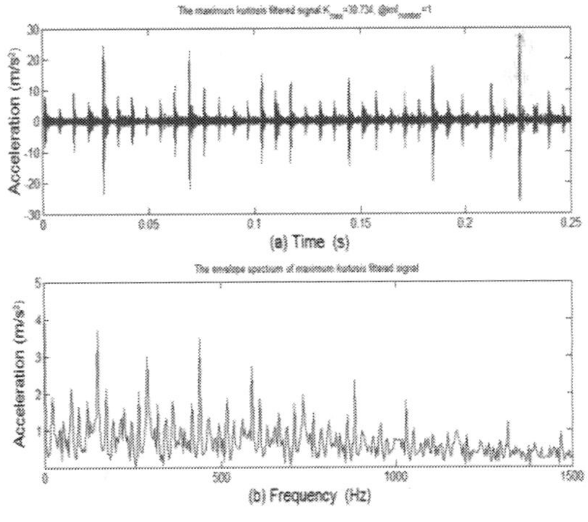

Fig. 5. The optimal kurtosis filtered data and its envelope spectrum.

In order to prove the effectiveness of the proposed method over the traditional envelope spectrum, Fig.4

shows the kurtosis of each IMF and the maximum kurtosis is in c_1 . Fig.5 shows the optimal kurtosis filtered signal c_1 and its envelope spectrum. In Fig. 5, one can clearly see the transient impact characteristics of the vibration data when the bearing inner race has defect. The fault of the bearing inner race leads to the sudden increase of vibration energy. For the faulty bearings, the transient impact feature caused by the interaction of roller defects can be clearly seen. In addition, these transient impacts reveal the failure mode of the bearing, corresponding to the bearing inner race characteristic fault frequency of 148.5Hz, which is caused by the structural defects of the inner race. This kind of repetition frequency reflects the deterioration of the inner raceway health when the bearing fault propagates through the bearing inner race.

4 Conclusions

In order to effectively extract the transient impact features of noisy vibration signals, an empirical mode decomposition method based on spectral kurtosis is proposed. This method has obvious advantages in extracting transient impact signal, which is often an important index of early defects of bearing system. Firstly, the vibration data is adaptively decomposed into several intrinsic modal functions by EMD, then the spectral kurtosis of each empirical modal function is calculated, and finally the envelope spectrum of the intrinsic mode function with the largest spectral kurtosis is calculated. The experimental results show that the EMD based on spectral kurtosis can effectively extract bearing faults.

Acknowledgment

This research is a part of the research that is sponsored by the scientific research fund of Tianjin University of Technology and Education (No. KYQD202003). At the same time, I am very glad to express my sincere thanks to the anonymous reviewers.

References

1. R.Rubini, U.Meneghetti, Mechanical Systems and Signal Processing **15**, 287-302 (2001)
2. D. Abboud, J. Antoni, S. Sieg-Zieba, et al, Mechanical Systems and Signal Processing **84**, 200-226 (2017)
3. N.E Huang, Z. Shen and S.R.Long et al, Proceeding of Royal Society London. Series A **454**, 903-995 (1998)
4. N.E.Huang, Z.Shen and S.R.Long, Annual Review of Fluid Mechanics **31**, 417-457 (1999)
5. H. Li, Y. Zhang, H. Zheng, Journal of Mechanical Science and Technology **20**, 1781-1789 (2006).
6. H. Li, Y. Zhang, H. Zheng, Journal of Mechanical Science and Technology **23**, 291-301 (2009)

MATEC Web of Conferences **327**, 03004 (2020) https://doi.org/10.1051/matecconf/202032703004
ICMIE 2020

Design of steering system of a buggy

D. Santana Sanchez, *and A.* Mostafa*

School of Environment, Science and Engineering, Southern Cross University,
Lismore NSW 2480, Australia.

Abstract. The present paper discusses the design analysis and limitations of the steering system of a buggy. Many geometrical and performance characteristics of the designed steering system were considered to address the kinematic constraints and load carrying capacity of the steering elements. Ackremann geometry approach was used to assess the limiting steering angle, while Lewis bending formula with the inclusion of dynamic effects was employed to characterise the flexural properties of the rack and pinion steering system. Analytical results were numerically verified using ABAQUS/Explicit nonlinear finite element (FE) package. Good agreement has been achieved between analytical and numerical results in predicting the flexural behaviour of the steering rack and pinion system.

1 Introduction

The steering system plays a very important role throughout the entire design of a vehicle, as it is the interface between the driver and vehicle. As torque is provided by the driver onto the steering wheel, it rotates the steering column and entire steering mechanism as well. This leads to turning of the rotational motion into linear that transmitted to the wheels via the connected tie rods. Apart from having this function, steering systems allow for driver effort to be lowered in order to facilitate steering and make it more efficient. The components that make up the steering system also absorb a fraction of the road shock before making contact with the driver [1]. As several components make up a steering system, the complexity or simplicity are dependent on the desired features of the automobile and its geometrical limitations. It is noteworthy that the intended goal of the designed steering system is to ensure ease of use, effectiveness of operation and reliability of the mechanical components that last longer with minimal maintenance. Therefore, the introduction of general rack and pinion design process with dynamics factors according to the recommendations of the American Gear Manufacturers (AGMA) provides reliable analytical results, while FEA was required to build insightful picture of the stress and strain distributions across the areas of interest. The design of the steering shaft was performed following the distortion energy theory to determine the precise size based on static and fatigue stress concentrations. In general, the steering system of a vehicle consists of tie-rod linkages and steering arms forming a trapezoidal shape. As the system moves linearly, it forms an approximation of a parallelogram which skews one side according to the desired steering angle [2]. Ackermann principle for steering geometry is introduced in the current system to differentiate between the faster displacement rate of the outer wheel compared with the inner one. Front wheel steering and constant speed are some noteworthy hypotheses that are included in Ackermann principle which may affect the study of the slippage between the outer and inner wheels [3]. These aforementioned procedures were followed in order to achieve a comprehensive evaluation of a simplified steering system for the proposed buggy, as to study its mechanical performance under static loading.

2. Design procedure of steering system

2.1 Ackermann percentage

Ackermann percentage can be calculated according to the position of the steering arms (knuckles or Pitmans arms). This is done with the use of Freudenstein's equation [4], which is based on the known angles of the steering arms as the desired turn radius is set at 3.5 m.

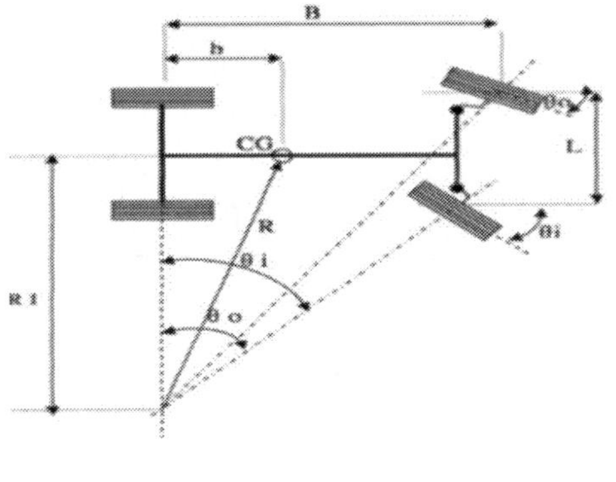

*Corresponding author: ahmed.thabet@scu.edu.au

 https://doi.org/10.1051/matecconf/202032703004

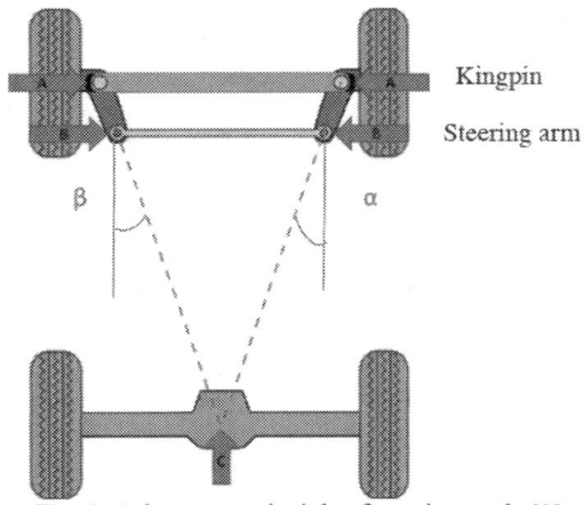

Fig. 1. Ackermann principle of steering angle [1].

Fig. 1 demonstrates the neutral position of the steering geometry, along with the kingpins and steering arms location. It can be seen that Ackermann's angle ∝ can be determined as:

$$\propto = tan^{-1}\left(\frac{\left(\frac{Sf}{2}\right)}{L}\right) \quad (1)$$

Where Sf is the distance between the kingpin and the centre of the tyre and L is the track-width.

2.2 Steering geometry

Fig. 1 shows the steering geometry required based on pro-Ackermann. The angles that each tyre is required can be determined based on the steer angle, hence rack travel can be determined.

$$R_1 = \sqrt{R^2 - B^2} \quad (2)$$

$$\theta_i = \tan^{-1}\left(\frac{B}{R_1 - \frac{L}{2}}\right) \quad (3)$$

$$\theta_o = \tan^{-1}\left(\frac{B}{R_1 + \frac{L}{2}}\right) \quad (4)$$

Where θ_o, θ_i, B, L, b, R_1 and R are the outside wheel turn angle, inside wheel turn angle, track width, wheel base, distance from rear axle to centre of mass, instantaneous centre of curvature and radius of turn, respectively. Therefore, the use of the greatest angle can be used in order to determine the steering ratio ($S.R$), as per the desired steering effort of 270°; ($S.R = 270/41.38 = 6.52$). Thus, for one complete turn of 160 mm steering wheel, the $S.R$ can be calculated using the following equation.

$$S.R = \frac{C}{(R.T)} \quad (5)$$

Where $S.R$ is the steering ratio based on desired steering effort and maximum angle needed to achieve a 3.5 m radius of turn, C is the circumference of the steering wheel and $R.T$ stands for rack travel.
In order to achieve centre to lock in both left and right directions, the calculated $R.T$ should be doubled to accommodate for rack travel in both sides.

2.3 Steering arm geometry

The distance from the centre of the steering wheel to the pinion (Fig. 2) was chosen as 0.1 m to fit with the rest of the buggy's components, as it would be best suitable for the allocated space for the driver.

Fig. 2. Geometry of the designed buggy.

The minimum distance between the rack and the king pin can be calculated from the geometry of the steering arm (Fig. 3) as per the following equations:

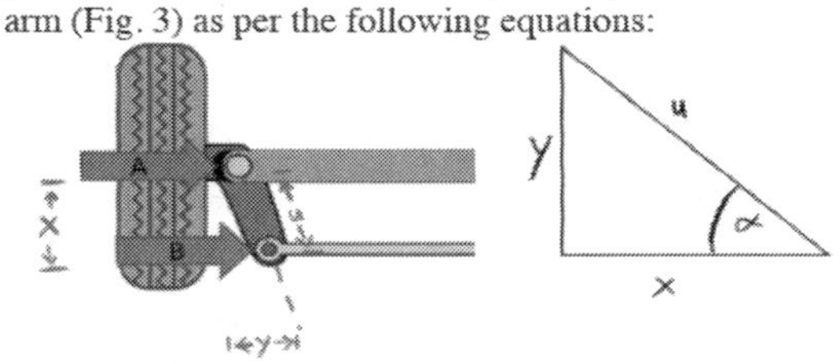

Fig. 3. Steering arm geometry.

$$u = \frac{x}{\cos(\propto)} \quad (6)$$

$$\sin(\propto) = \frac{y}{u} \quad (7)$$

Where u, y, x and $\propto$ are the steering arm length, arm base, distance away from king pin and Ackermann angle, respectively. The length of the overall tie rod can be determined by subtracting the arm base from the distance between the kingpins.

$$Lrod = Sf - 2y \quad (8)$$

This distance must include the rack and tie rods. By employing the sum of moments equation and taking into account the static frictional force within a tyre, the force needed to turn the tyre at its greatest effort can be determined (in the current design, a lateral force of 201.354 N is needed).

3. Analytical and numerical modelling of steering components

3.1 Analytical model of the rack and pinion

The spur gear analysis was performed based on the amount of steering desired and the force that the gear is intended to transmit. Hence determining the pitch diameter *(Dp)* of the gear can be calculated based on the known requirements, as the gear is required to travel the rack from centre to lock in within 270°.

$$Dp = \frac{\left(2 * R.T * \frac{3}{4}\right)}{\pi} \quad (9)$$

Followed by the number of teeth being determined:

$$N = \frac{Dp}{m} \quad (10)$$

Where N and m are the number of teeth and module of the gear, respectively. Hence torque can be found as:

$$T = FL * rp \quad (11)$$

Where FL and rp are the lateral force and the radius of the pinion, respectively. This allows for the force at the steering wheel to be determined,

$$T = Fst * rst \quad (12)$$

Where Fst and rst are the lateral force at steering wheel and the radius of the steering wheel, respectively. The dimensions of the rack was based on the dimensions of the pinion, thus a 3 mm module was adopted in order to minimize the number of teeth yet increasing the diameter of the pinion which lowered the lateral force.
By utilizing the AGMA stress equation(σ) and the Lewis bending equation [5], various dynamic factors can be considered and compared with the numerical results to determine the reliability and accuracy of the design. AGMA stress formula is given as:

$$\sigma = \mathrm{W^T K_o K_v K_s} \left(\frac{1}{\mathrm{bm_t}}\right)\left(\frac{\mathrm{K_H K_B}}{\mathrm{Y_J}}\right) \quad (13)$$

Where W^T, K_o, K_v, K_s, b, m_t, K_H, K_B and Y_J are the tangential transmitted load (N), the overload factor, the dynamic factor, size correlation factor, the face width of the narrower member, transverse metric module, load-distribution factor, rim-thickness factor and geometry factor for bending, respectively [3]. This would later allow for factors of safety of wear and bending to be concluded. In order to estimate the flexural strength of the pinion's teeth, Lewis bending formula (σ_{LB}) was used (Equation 15). The formula allows for the determination of the teeth bending strength taking into consideration the dynamic characteristics of the applied load.

$$\sigma_{LB} = \frac{K_v W}{FmY} \quad (14)$$

Where K_v, W, F, m and Y are the dynamic factor, tangential load (N), face width, module and bending factor, respectively.

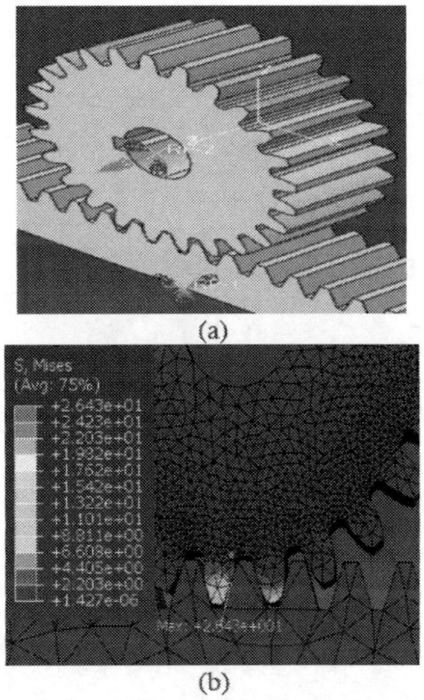

(a)

(b)

Fig. 4. Numerical modelling of the rack and pinion (a) boundary conditions and (b) Von Mises stress distribution.

3.2 Numerical Model of the Rack and Pinion

FE models were generated using ABAQUS software [6] to simulate the loading cases between the rack and pinion of the steering system. Both parts were modelled as linear elastic material with elastic modulus and Poisson's ratio equal 205 GPa and 0.29, respectively. 3D continuum element C3D10 was used to represent the geometry of the simulated parts to capture the stress concentration at the mating surfaces. Penalty contact interaction with 0.1 friction coefficient was used to model the tangential contact behaviour, while hard contact was utilized to prevent any pressure-overclosure at the interface. Based on the geometry, a total of 92209 quadratic tetrahedral elements were used with bias mesh distribution in the vicinity of the tooth face to capture the stress concentration in the pinion. The pinion was constrained from moving in all directions except the rotation around the z-axis, while the rack was absolutely fixed to simulate the ultimate resisting load as shown in Fig. 4 (a).

Fig. 4 (b) shows the numerical distribution of the Von Mises stress with a maximum value of 26.43 MPa at the tooth's face. The calculated flexural strength based on Lewis formula was 25.77 MPa, provided the average directional stress not the effective stress (Von Mises). This difference between the flexural strength based on Lewis formula and the effective stress could be attributed to the assumptions made in deriving Lewis equation such as the gear teeth are cantilever beam with load applied at the end of the tooth, radial forces are neglected and various factors such as sliding friction and stress concentrations based on changes of the area are neglected [5].

4. Conclusions

Analytical and numerical approaches were used to determine the reliability and the design constraints of the steering system of a buggy. Steering angle, ratio and geometry were estimated based on Ackermann principle to evaluate the kinematic constraints of the designed buggy. FE results based on static loading conditions have shown safe working conditions under selected design parameters. The employed rack and pinion have shown remarkable resistance to the applied bending load which ensures sufficient safety factor. Moreover, the analytical analysis using AGMA equation upon both rack and pinion provided results in close agreement with the FE ones.

References

1. A.; WJ, J. WG. Steering System Principles. Popular Mechanics; 2014.
2. Qiu Q, Fan Z, Meng Z, Zhang Q, Cong Y, Li B, et al. Extended Ackerman Steering Principle for the coordinated movement control of a four wheel drive agricultural mobile robot. Computers and Electronics in Agriculture. 2018;152:40-50.
3. Budynas RG, Nisbett JK, Shigley JE. Shigley's mechanical engineering design2015.
4. Ghosal A. The Freudenstein Equation. Design of Fourl-Link Mechanisms. 2010.
5. Budynas RG, Nisbett JK. Mechanical Engineering Design. New York: Mc Graw HIll Education; 2015.
6. ABAQUS. ABAQUS 2018 EF Documentation. 2018 ed: Simulia; 2018. p. Finite Element Analysis.

Mobile robot positioning algorithm based on Kalman filtering method in network environment

Shuang Zhang [1,*]

[1] Department of Data Science and Technology Heilongjiang University, Harbin150080, Heilongjiang, China

Abstract. Positioning is the basic link in a multi-mobile robot control system, and is also a problem that must be solved before completing a specified task. The positioning method can be generally divided into relative positioning and absolute positioning. Absolute positioning method refers to that the robot calculates its current position by acquiring the reference information of some known positions in the outside world, calculating the relationship between itself and the reference information. Absolute positioning generally adopts methods based on beacons, environment map matching, and visual positioning. The relative positioning method mainly uses the inertial navigation system INS. The inertial navigation system directly fixes the inertial measurement unit composed of the gyroscope and the accelerometer to the target device, and uses the inertial devices such as the gyroscope and the accelerometer to measure the triaxial angular velocity and The three-axis acceleration information is measured and integrated, and the mobile robot coordinates are updated in real time. Combined with the initial inertial information of the target device, navigation information such as the attitude, speed, and position of the target device is obtained through integral operation [1-2]. The inertial navigation system does not depend on external information when it is working, and is not easily damaged by interference. As an autonomous navigation system, it has the advantages of high data update rate and high short-term positioning accuracy [3]. However, under the long-term operation of inertial navigation, due to the cumulative error of integration, the positioning accuracy is seriously degraded, so it is necessary to seek an external positioning method to correct its position information [4]

1 Introduction

In order to solve the problem of inertial navigation under long-term operation due to the cumulative error of integration, the positioning accuracy is reduced and GPS absolute positioning cannot be used. This paper proposes a WSN / INS combined positioning algorithm combined with wireless sensor network positioning technology. The algorithm first constructs the coordinate system for the mobile robot working environment, provides a working environment for establishing the WSN / INS combined positioning system, and secondly combines the conversion methods of the combined positioning system under different coordinate systems, and uses the motion characteristics of WMR to establish the combined positioning system. The equation of state, and finally according to the linear characteristics of the established state space model, the Kalman filter is used to combine the data. The WSN / INS navigation method uses the sensor data of INS and WSN, and calculates the redundant or complementary information of these data in space or time through the Kalman filter according to the specified method to obtain unified and consistent positioning information. Because the data collected by multi-sensor nodes is redundant and complementary, it can expand the observation scale in time and space, and strengthen the credibility of effective data. This is more robust and effective for mobile robot formation systems in dynamic environments. Reliability and improving working accuracy are very important. The research work in this chapter will provide a theoretical basis for the subsequent navigation and formation control research

2 MOBILE ROBOT CONTROL FRAMEWORK UNDER NETWORK ENVIRONMENT

In the sensor network environment, the control requirements of mobile robots show new characteristics compared with the past [5-7]: In the network environment, mobile robots can perceive information about obstacles in the workspace through WSN. The spatial characteristics of sensor network information are often affected by communication parameters. At the same time, WSN has the characteristics of strong awareness and diversified information resources. In WSN, environment awareness information is obtained through nodes, and awareness information is also stored on the nodes. This poses new challenges to the placement density of nodes, the selection of effective data, and the effectiveness of information dissemination. Therefore, when WMR, as the main performer,

* Corresponding author: 2006118@hlju.edu.cn

combines sensor networks, it is no longer only a participant in network communication, but also a provider of partial information.

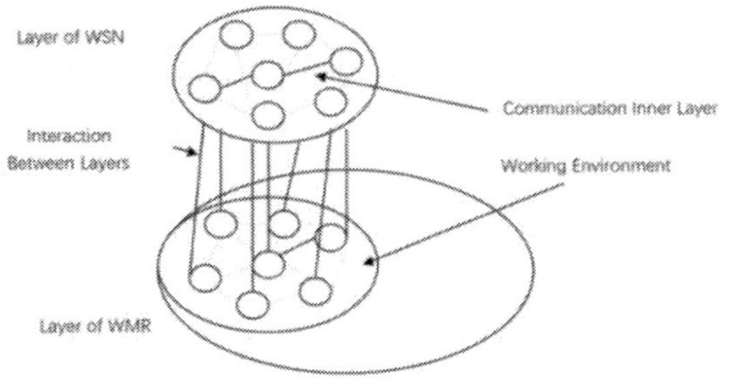

Fig. 1. Mobile formation control method framework based on WSN network.

3 WSN / INS combined positioning method under loose coupling

In the combined design of wireless sensor networks and inertial navigation equipment systems, loosely coupled systems are defined as systems in which each component in the WSN / INS system has little or no knowledge of other individual components. The goal of loose coupling is to minimize the dependence of INS on external navigation equipment. The external navigation equipment in the WSN / INS system can be replaced by other similar functional components.

The proposed WSN / INS combined positioning method first constructs the coordinate system for the WMR working environment, and provides a working environment for establishing the WSN / INS combined positioning system. Combining the transformation methods of different coordinate systems of the combined positioning system, the motion equation of the WMR is used to establish the state equation of the combined positioning system. According to the linear characteristics of the established state space model, the Kalman filter is used to combine the data.

The expression of time continuous navigation in the global coordinate system is

$$\begin{cases} \dot{r^e} = v^e \\ \dot{v^e} = R_b^e f^b - 2\Omega_{ie}^e v^e + g^e \\ \dot{R_b^e} = R_b^e \Omega_{eb}^b \end{cases} \quad (1)$$

r^e and v^e represents the position and speed in the respective coordinate system. f^b is the acceleration value measured under the mobile robot coordinate system. The angular marks e, b used to indicate which coordinate system in the current navigation system. They are the earth coordinate system, mobile robot coordinates and inertial gyroscope coordinate system. When the acceleration of the earth is known in the global coordinate system, g^e is the position compensation value. R_b^e is a rotation matrix, which can convert the vectors in the mobile robot coordinate system to the global coordinate system. R_b^e has nine elements and it has three degrees of freedom, so R_b^e can be uniquely described by three angular velocities. The matrices Ω_{eb}^b and Ω_{ie}^e are skewed symmetric matrices, which represent the angular rates of ω_{eb}^b and ω_{ie}^e.

$$\begin{cases} a \times \omega_{eb}^b = \Omega_{eb}^b a \\ a \times \omega_{ie}^e = \Omega_{ie}^e a \end{cases} \quad (2)$$

In formular（2）Where a is a 3 × 1 three-dimensional vector, and the matrix Ω_{eb}^b is expressed as follows：

$$\Omega_{eb}^b = \begin{pmatrix} 0 & -\omega_{eb_z}^b & \omega_{eb_y}^b \\ \omega_{eb_z}^b & 0 & -\omega_{eb_x}^b \\ -\omega_{eb_y}^b & \omega_{eb_x}^b & 0 \end{pmatrix} \quad (3)$$

In equation (3), the elements $\omega_{eb_x}^b$ $\omega_{eb_y}^b \omega_{eb_z}^b$ are the angular velocities of the robot coordinate system relative to the earth coordinate system. Solve in the robot coordinate system. The matrix Ω_{ie}^e e is expressed as follows

$$\Omega_{eb}^b = \begin{pmatrix} 0 & -\omega_{eb_z}^b & \omega_{eb_y}^b \\ \omega_{eb_z}^b & 0 & -\omega_{eb_x}^b \\ -\omega_{eb_y}^b & \omega_{eb_x}^b & 0 \end{pmatrix} \quad (4)$$

The input of INS is a three-dimensional value, which is derived from the acceleration measured under the robot coordinate system with respect to the reference value ω_{ib}^b in the inertial coordinate system.

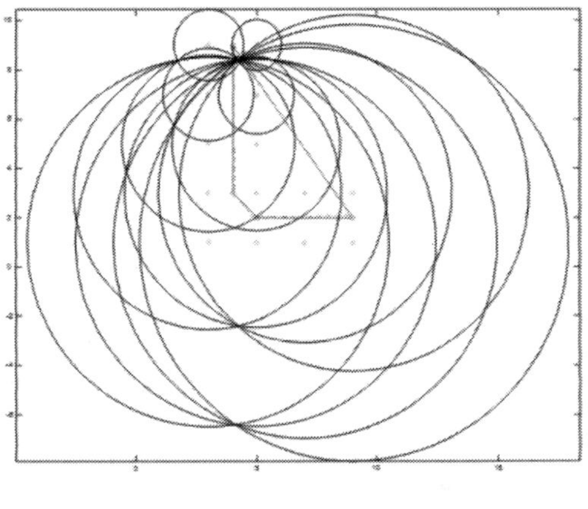

a) trajectory 1

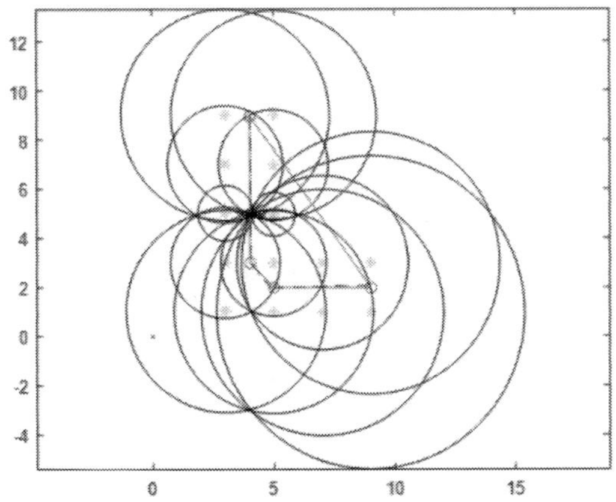

b) trajectory 2

Fig. 2. WSN/INS schematic diagram of navigation trajectory under 14 positioning points

Figures a and b show the comparison between the actual trajectory and the theoretical trajectory of the WSN / INS navigation system when WSN provides 14, 26 anchor nodes for the INS navigation system. The solid black circle is the working area of each anchor node. In the experimental simulation process, WMR takes [4,9] coordinates as the starting point and runs to the end point coordinates [9,2]. As can be seen from Figure 3, the deviation of the WMR actual path (blue dotted line) and the WMR theoretical path (red solid line) becomes smaller as the number of WSN anchor nodes (light green solid points) increases.

Table 1. Comparison of position root mean square error of different positioning methods

Positioning method	East direction error (mm)
INS	1.2
WSN	0.62
WSN/INS filter	0.51
Positioning method	1.2

4 Conclusion

The experimental results show that the WSN / INS integrated navigation system and integrated navigation algorithm designed in this paper are feasible and effective in positioning mobile robots in actual scenarios. Experimental results show that, in terms of network connectivity, under different communication radius conditions, Compared with INS and WSN respectively positioning WMR, the proposed WSN / INS position solution method has a comprehensive positioning accuracy 18% and 59% higher than that of WSN and INS respectively; The goal also lays a theoretical foundation for the establishment of a navigation and formation control system under the network environment.

5 Acknowledgement

This work is supported by 2019 Heilongjiang universities basic scientific research business fee special fund project of Heilongjiang University (NO. KJCX201922).

References

1. Nepali M M R, Yadav M N, Balasubramaniah M S, et al. A Novel Wall Following Algorithm for Mobile Robots[J]. Computer Science Journals, 2014.
2. Parkinson B W. Global Positioning System: Theory and Applications, Volume II (AIAA)[J]. Aiaa, 1996, 163.
3. Qian C, Liu H, Tang J, et al. An Integrated GNSS/INS/LiDAR-SLAM Positioning Method for Highly Accurate Forest Stem Mapping[J]. Remote Sensing, 2016, 9(1):3.
4. Das A K, Fierro R, Kumar V, et al. A vision-based formation control framework[J]. IEEE Transactions on Robotics & Automation, 2002, 18(5):813-825.
5. These two Army BCTs will be the first to put robotic vehicles in their formations. Todd South. [R/OL].2018. www.armytimes.com
6. Internet of Things: Wireless Sensor Networks[R/OL]. White paper. International Electrotechincal Commission.2014.www.iec.ch
7. Deng X P, Liu Z, Jiang W L, et al. Passive location method and accuracy analysis with phase difference rate measurements[J]. Radar, Sonar and Navigation, IEE Proceedings -, 2001, 148(5):302-307.

Kinematic Performance Analysis of a Controllable Mechanism Welding Robot with Joint Clearance

Junjie Gong[1], *Wei* Wei[1,2], *Ganwei* Cai[1*], *Yixin* Liu[1] and *Sixu* Peng[1]

[1]School of Mechanical Engineering, Guangxi University, Nanning, China
[2]School of Mechanical and Automotive Engineering, South China University of Technology, Guangzhou, China.

Abstract. Because the motor and reducer of the conventional tandem welding robot are installed at the revolute joint, the robot has large rotational inertia and long residual vibration time. However, due to the limitation of workspace and other reasons, the current parallel robot is not suitable for all series welding robots. Therefore, based on the concept of "multi-degree-of-freedom controllable mechanism", a new controllable mechanism welding robot is proposed in this paper. The driving motor and reducer, which have great influence on the main and branch chain of the welding robot, are installed on the frame. The advantage of this design is that the inertia of the robot mechanism is significantly reduced, and its dynamic performance is improved. The position errors of the end members of the welding robot moving along the circle, as well as the angle errors of each joints under the circle movements are analysed.

1 Introduction

As a common industrial robot, welding robot is widely used in production practice 1. Because the driving motor and reducer are installed in the joint of the traditional series welding robot, not only the moment of inertia of the robot is increased, but also the residual vibration is difficult to eliminate. However, the current parallel robot has some problems, such as limited working space and high coupling degree, so it is difficult to replace the traditional series welding robot.

Due to the accumulation of the clearance value of each kinematic pair of the traditional series robot, the position error of end components is large. Somashekhar et al.2 proposed a robot vision measurement system, which is used to obtain the motion error of the end position of the mobile robot.Tsai et al. 3used the equivalent kinematic pair to simulate the degree of freedom caused by the joint clearance, and studied the influence of the clearance error of the kinematic pair on the transmission performance of the mechanism. Shi et al. 4 improved the absolute positioning accuracy of KUKA KR5 arc series industrial robot with joint clearance by using the method of on-line error compensation. Zhou et al. 5improved a hybrid least-squares genetic algorithm-based algorithm, and identified and compensated the precision of HA006 series robot with the clearance of joint.

In this paper, the concept of "multi degree of freedom controllable mechanism"6 is applied to industrial robots, and a welding robot with multi degree of freedom controllable mechanism is proposed7. The basic idea of this new type of multi DOF welding robot is to move the motor and reducer installed in the motion joint of the serial robot to the frame as much as possible. The motor is installed on the frame to drive the motion chain, so as to realize the movement of the robot. In this way, the inertia of the robot can be obviously reduced and the time of residual vibration can be shortened. However, the kinematic performance of this new type of multi-degree-of-freedom robot has not been analyzed and studied, especially the influence of its motion pair clearance on its end trajectory. Therefore, based on the clearance model of the continuous contact kinematic pair8, the D-H model9 with clearance of the main kinematic chain of the welding robot is established, and the position error of the end member and the angle error of each joint are analyzed when there is joint clearance under the arc trajectory of the welding robot, which lays the foundation for the follow-up research of the controllable mechanism welding robot.

2 Structure of welding robot with controllable mechanism

The controllable mechanism welding robot transfers part of the joint drive motor and reducer to the base, reduces the moment of inertia and residual vibration of the robot, and improves the dynamic performance of the robot. The structure of controllable mechanism welding robot is shown in Figure 1.

* Corresponding author: caiganwei@163.com

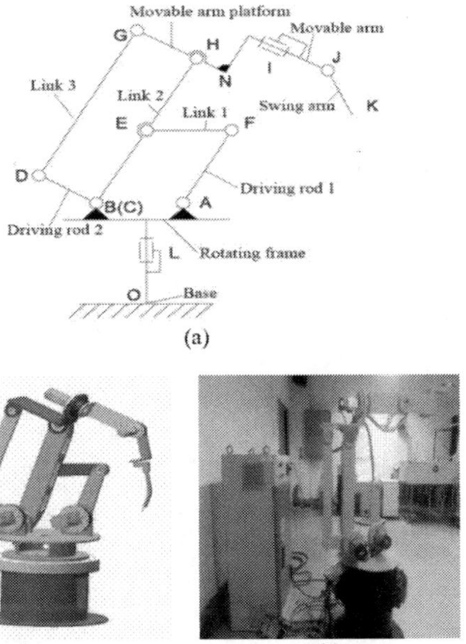

(b) (c)

Figure 1(a) Mechanism diagram of welding robot. **(b)**A three-dimensional picture of a welding robot. **(c)**The physical prototype of the welding robot.

3 Kinematics analysis of welding robot considering joint clearance

3.1. Vector model of joint clearance

Establish the clearance model of the kinematic pair as shown in Figure 2. It is assumed that the holes and shafts of the rotating pair are all standard circles, and the radii are R_b and R_J respectively. When the clearance is not considered, the ideal length of member A is l, and the ideal angular displacement is θ. When considering the clearance between moving pairs, the actual length of member a is l', and the actual angular displacement is θ'. The following relationships can be obtained:

$$\begin{cases} l' = \sqrt{(l + r\cos\varphi)^2 + (r\sin\varphi)^2} \\ \theta' = \theta + \arcsin\dfrac{r\sin\varphi}{\sqrt{(l + r\cos\varphi)^2 + (r\sin\varphi)^2}} \end{cases} \quad (1)$$

According to the actual working track of the controllable welding robot studied in this paper, the clearance model of the moving pair can be regarded as a continuous contact model, and the clearance of the moving pair can be regarded as a certain length of connecting rod without mass. According to the error model:

$$r = R_B - R_J, \varphi \in [0, 2\pi] \quad (2)$$

3.2. Forward kinematics solution of welding robot considering joint clearance

Each component of the welding robot is connected by a moving pair. This paper only studies the influence of joint clearances 2 and 3 on the robot end position. The reasons are as follows: a. the joints 1 and 2 are located on the main chain of motion, and the main motion of the whole robot is completed by the main chain of motion; B. the joints 2 and 3 are far away from the end components of the robot, where the clearance has a greater impact on the end through the accumulation of the chain of motion. Because the mechanism welding robot is a complex structure of series parallel combination, its clearance model can be regarded as a continuous contact model to simplify the analysis. Based on this, the kinematics model of welding robot with clearance is established, as shown in Figure 3.

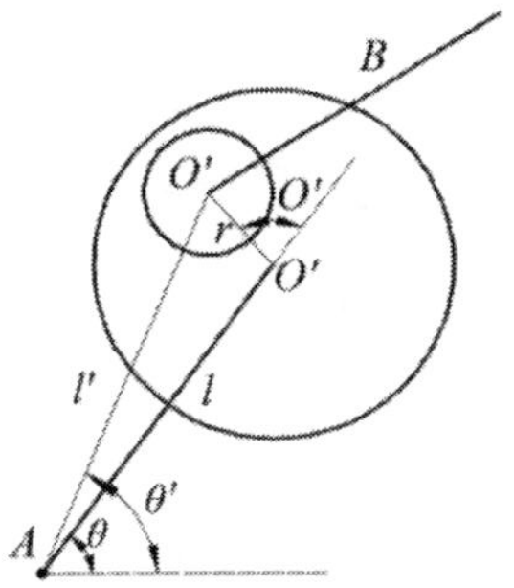

Figure 2 Model of joint clearance

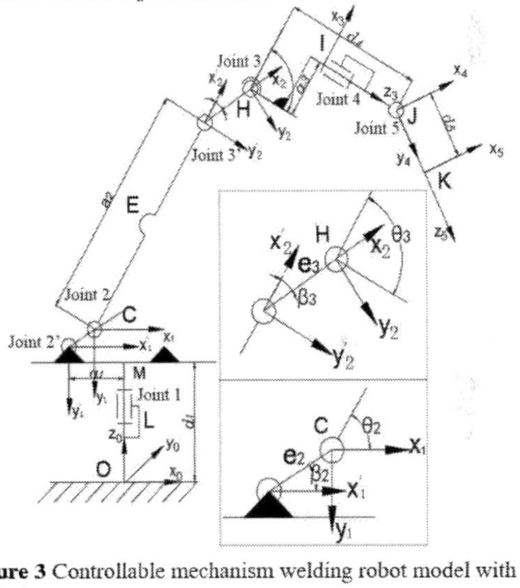

Figure 3 Controllable mechanism welding robot model with joint clearance

The synthetic transformation matrix of robot end is

$$^{0}_{5}T = \prod_{j=1}^{5}\left(^{j-1}_{j}T(\theta_{j-1})\right) = {}^{0}_{1}T \cdot {}^{1}_{2}T \cdot {}^{2}_{3}T \cdot {}^{3}_{4}T \cdot {}^{4}_{5}T \quad (3)$$

According to formula(3), the position transformation matrix of welding robot with clearance joint is as follows

$$^{0}_{5}T' = \begin{bmatrix} R_{3\times3} & P \\ 000 & 1 \end{bmatrix} = {}^{0}_{1}T\, {}^{1}_{2'}T\, {}^{2'}_{2}T\, {}^{2}_{3'}T\, {}^{3'}_{3}T\, {}^{3}_{4}T\, {}^{4}_{5}T = \begin{bmatrix} n_x' & o_x' & a_x' & p_x' \\ n_y' & o_y' & a_y' & p_y' \\ n_z' & o_z' & a_z' & p_z' \\ 0 & 0 & 0 & 1 \end{bmatrix} \quad (4)$$

Where, p_i' is the position matrix of the end actuator of the welding robot with clearance in the base coordinate; $\begin{bmatrix} n_i' & o_i' & a_i' \end{bmatrix}$ is the position matrix of the end actuator in the base coordinate.

4 Analysis of the position error of the end component of welding robot

4.1 Analysis of motion track and joint angle

The arc trajectory of the welding robot is simulated, and the trajectory equation is

$$x_0 = 68 - 30\sin(0.5\pi t), y_0 = 0, z_0 = 85\text{-}30\cos(0.5\pi t) \tag{5}$$

Substituting Eq. (5) into Eq.(4), draw the arc track with Matlab, and then get the speed change of each joint of the clearance welding robot and the ideal welding robot, as shown in Fig. 4.

The arc track motion of welding robot is analyzed. In an ideal state, the trajectory of the robot's end and the angle of its moving joint change smoothly. When considering the clearance, the angle of the moving joint fluctuates slightly. Therefore, in the actual process, due to the reasonable setting of the clearance size of the motion pairs.

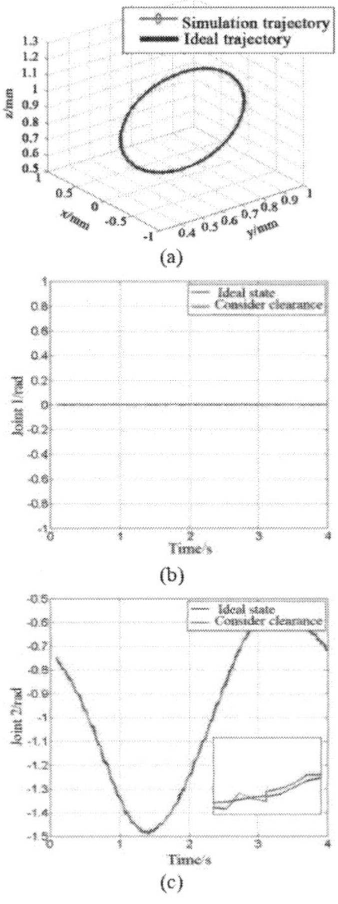

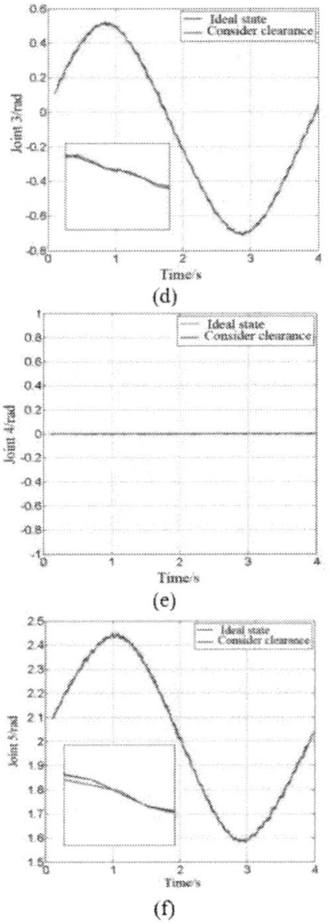

Figure 4 (a) Arc trajectory simulation of welding robot. **(b)-(f)** Angle error of joint 1-joint 5

4.2 Analysis of position error of end component

According to the kinematic analysis of the welding robot, the difference between the positive kinematics of the ideal model and that of the joint clearance model can be obtained

$$\begin{cases} \Delta_x = e_2 c\beta_2 c\theta_1 + e_3 c(\beta_3 + \theta_2) c\theta_1 \\ \Delta_y = e_2 c\beta_2 s\theta_1 + e_3 c(\beta_3 + \theta_2) s\theta_1 \\ \Delta_z = -e_2 s\beta_2 - e_3 c(\theta_2 + \theta_3) \end{cases} \tag{6}$$

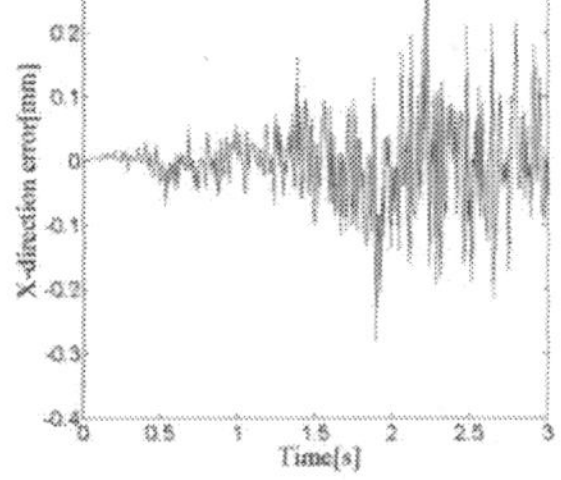

Figure 5 X-direction positio-nal value error of end compon-ent

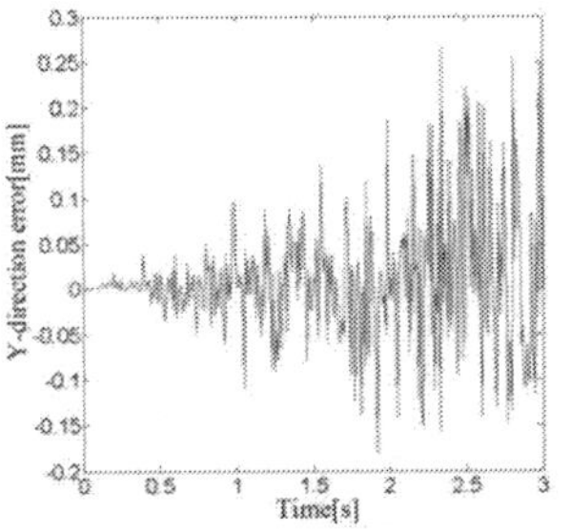

Figure 6 Y-direction positio-nal value error of end compon-ent

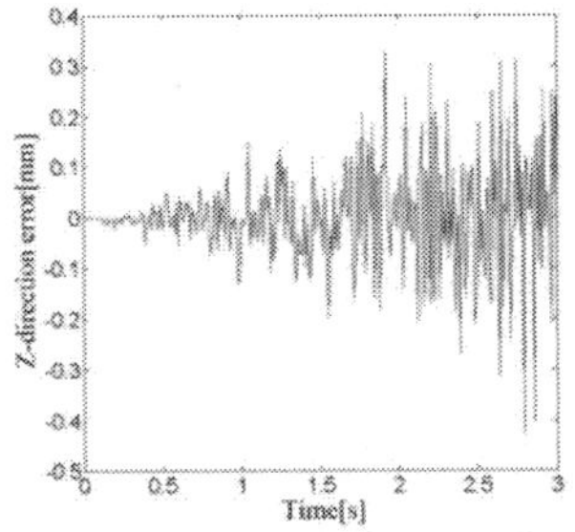

Figure 7 Z-direction posi-tional value error of end co-mponent

It can be seen from equation (6) that the position deviation of the robot's end and the eccentricity angles β_2, β_3 and the eccentricity distances e_2 and e_3 are positive cosine relations, which are difficult to measure and are in chaotic motion. Therefore, it can be taken that the eccentricity angle obeys the uniform distribution $U(0,2\pi)$ and the eccentricity obeys the normal distribution $N(0.1,0.1)$. Set the clearance value to *0.1mm*, in 3 seconds, θ_1 from 0° to 60°, θ_2 and θ_3 from -60° to 0°. The variation rule of robot end component error is shown in Figure 5 to Figure 7.

According to figure 8, when the clearance value is 0.1mm, the error range of the robot in X direction is (0-0.3mm), in Y direction is (0-0.25mm), and in Z direction is (0-0.5mm). It can be seen that when the clearance value is 0.1mm, the maximum error of the end position is 0.5mm, which has a great influence on the quality of the welding machine required for high-precision welding. Therefore, it is the key to design the welding robot to set the allowable clearance value reasonably.

5 Conclusion

In this paper, a new type of multi-DOF robot mechanism is proposed. The part of motors and reducers that were originally installed in the motion joints are installed on the frame, which reduces the inertia of the robot and improves the dynamic performance of the robot. In this paper, its kinematics performance is studied. The D-H method is used to establish the kinematic model of the welding robot with clearance. Based on the clearance model of continuous contact, the angle variation relationship of each joint of the welding robot with clearance is analyzed. When the clearance value is 0.1mm, the range of end position error of welding robot is obtained.

The author(s) disclosed receipt of the following financial support for the research, authorship, and/or publication of this article. The work was supported by National Natural Science Foundation of China Grant No. 51765005.

References

1. Haitao L, Jia F, Lichuang J, Guangming L, Chuangshuai Y, and Tingke W. Kinematics and dynamics analysis of a new-type friction stir welding robot and its simulation. *Advances in Mechanical Engineering*, vol. 11, pp.1-12, 2019.
2. Somashekhar S Hiremath, Robins Mathew, and Jennifer Jacob. Implementation of Low Cost Vision Based Measurement System: Motion Analysis of Indoor Robot, *International Journal of Mechanical Engineering and Robotics Research*, Vol. 7(6), pp. 575-582, 2018.
3. Tsai M J, Lai T H. Kinematic sensitivity analysis of linkage with joint clearance based on transmission quality. *Mechanism and Machine Theory*,vol. 39,pp.1189-1206,2004.
4. Xiaojia S, Fumin Z, Xinghua Q, Bailing L,and Junlong W. Position and Attitude Measurement and Online Errors Compensation for KUKA Industrial Robots. *Journal of Mechanical Engineering*, 2017, 53(8): 1-7.
5. Jian Z, Heejun K. A hybrid least-squares genetic algorithm-based algorithm for simultaneous identification of geometric and compliance errors in industrial robots. *Advances in Mechanical Engineering*, vol. 7(6),pp.1-12,2015.
6. Ganwei C, Yuchen P, Hongzhou W, and Jiansheng D. Functional Analysis Based Type Synthesis of a Novel Type of Loading Mechanisms. *Journal of Mechanical Engineering*.vol. 50,pp.50-59,2014.
7. Ganwei C, Mingyu D, Huali R, Jianbo T, Chengliang W, Honghan L, and Bin Y. Five-degree-of-freedom series-parallel controllable mechanism welding robot, China, Patent, 201610464088.1 [P].2016-06-23.
8. Flores P, Ambrosio J, Claro JCP, Lankarani HM, Koshy CS. A study on dynamics of mechanical systems including joints with clearance and lubrication. *Mechanism & Machine Theory*, vol, 41, pp. 247-261, 2006.
9. Xu Z, Zelong Z, Yong Q. Parameter Identification and Calibration of D-H Model for 6-DOF Serial Robots. *ROBOT*, vol, 38,pp. 360-370, 2016.

MATEC Web of Conferences **327**, 03007 (2020) https://doi.org/10.1051/matecconf/202032703007
ICMIE 2020

A risk control framework for safe manufacturing workstations in view of COVID-19

Ebly Sanchez[1,*] *and Knut Åkesson*[2]

[1]Volvo Group Truck Operations, Greensboro, USA
[2]Department of Electrical Engineering, Chalmers University of Technology, Sweden

Abstract. The manufacturing industry resumes operations after the COVID-19 pandemic supported by return-to-work guidelines, which are mostly personal protection measures for the workers and employees. In this paper, we propose a framework for assessing risk at the workstation level by linking the risk levels to possible mitigation strategies that can be implemented using standard operating procedures (SOP), 5S and problem-solving. Within industrial plants, operators work in close contact with coworkers and supervisors, and they are also sharing tools and machines. It is therefore, essential to develop strategies that reduce the operator's exposure to viruses in the workplace. The purpose of this work is that when implemented, the risk assessment model and specifically how SOP, 5S and problem solving can be used to implement administrative and engineering controls resulting in a safe workplace and increasing level of confidence for the operators working within the plant.

1 Introduction

Many uncertainties remain concerning when the COVID-19 coronavirus will reach its peak in every country and globally. At present, the therapeutic strategies to deal with the infection are only supportive, and prevention aimed at reducing transmission in the community is our best weapon [2]. It is clear that COVID-19 has disrupted operations on a massive scale, challenging manufacturers to assess the impact continually and, in the best way, respond quickly and preventively. The manufacturing industry is resuming operations supported by return-to-work guidelines [3–5, 9], which is providing the needed virus protection for workers and employees. Because COVID-19 cannot be eliminated and is likely to be around for a long time, it is highly necessary to increase the safety and trust of the workers and employees working at manufacturing sites. Therefore, managers need guidance on how to motivate the trust of the workers and at the same time secure the daily production process with the safety of the employees at the highest priority. Resuming operations while developing and implementing COVID-19 prevention measures has been not only a challenge but also a massive learning process where new knowledge is being developed regarding more proactive steps to avoid virus transmission in the future.

The risk of disease transmission of COVID-19 depends on several factors. Some of these factors are described in the U.S. Department of Labor and U.S. Department of and Health and Human Services' [2] which include:

- Distance between workers. Manufacturing workers often work close to one another on production or assembly lines. Workers may also be near one another at other times, such as when clocking in or out, during breaks, or in locker/changing rooms.
- Duration of contact. Manufacturing workers often have prolonged closeness to coworkers (e.g., for 8–12 hours per shift). Continued contact with potentially infectious individuals increases the risk of COVID-19 transmission.
- Type of contact. Manufacturing workers may be exposed to the infectious virus through respiratory droplets in the air — for example, when workers in a plant have the virus cough or sneeze. It is also possible that exposure could occur from contact with contaminated surfaces or objects, such as tools, workstations, or break room tables. Shared spaces such as break rooms, locker rooms, and entrances/exits to the facility may contribute to their risk.

The main contribution in this paper is a framework that combines existing approaches to risk assessment, risk control approaches, standard operating procedures to implement risk control, 5S, and problem-solving for a sustained improvement cycle. We provide an example of how to the risk assessment together with the evaluation of different risk control approaches to reduce risk levels. Also, an example of how a standard operating procedure can be used to implement an administrative risk control approach at the workstation level is presented.

The outline of the paper is that the different techniques are briefly introduced in section 2. The proposed model is presented in Section 3. In Section 4 it is discussed how the proposed model can be continuously updated and reviewed.

2 Background

In this section, we give an introduction to the hierarchy of controls that categorize possible means to handle hazards.

2.1 Risk analysis

Risk analysis, [1], is routinely used in the manufacturing industry to assess the risk for physical accidents at the workstation. However, the risk analysis in the manufacturing industry has not traditionally included potential exposure to viruses. The risk priority number (RPN), which is the product of the severity, occurrence, and detection,

* Corresponding author: ebly.sanchez@volvo.com

https://doi.org/10.1051/matecconf/202032703007

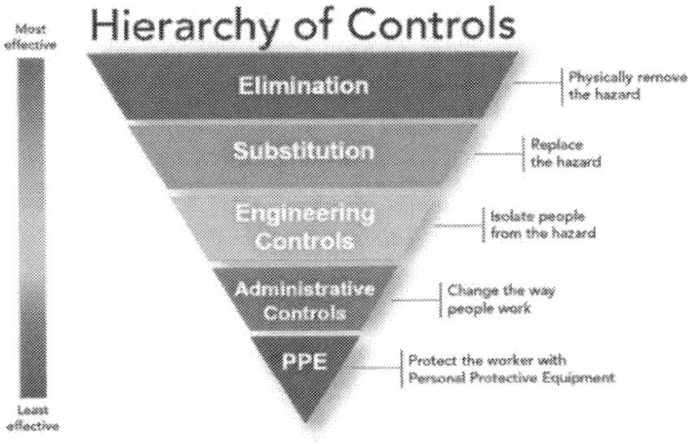

Fig. 1. Hierarchy of Controls (HOC), from [10].

is used to prioritize potential risks against each other to focus on those risks that have the most considerable consequences. In this work, we introduce a valid and straightforward approach to risk analysis similar to the RPN approach but adapted to handle risk related to exposure to the virus.

2.2 Hierarchy of controls (HOC)

Controlling exposures to occupational hazards is the fundamental method of protecting workers. Traditionally, a hierarchy of controls, see Fig. 1, has been used as a means of determining how to implement feasible and effective control solutions. Risk controls were categorized using a modified version of the National Institute for Occupational Safety and Health (NIOSH) hierarchy of risk controls [10].

Elimination Stop using the hazardous process or materials.

Substitution Substitute a less hazardous process or material.

Engineering controls With a focus on physical barriers, isolation, forcing functions, human factors, and fail-safe design, engineering controls improve safety independent of worker interactions.

Administrative controls Policies, procedures, training, and other controls that depend on people taking the correct actions.

Personal protective equipment (PPE) such as face-masks, face-shields, safety glasses, and gloves.

Elimination and substitution are the most effective approaches, while PPE is typically considered the least effective approach. In this work, we mainly focus on what can be done at the engineering control and the administrative controls level to reduce risk at the workstation.

2.3 Standard Operating Procedures (SOP)

SOP [7] has become the base for improvement in the industry with clear benefits. In the SOPs, it is precisely described how to perform the job, i.e., it describes the step sequence. The use of SOPs typically leads to a reduction of variation and error correction, improved safety, facilitates communication, and increasing labor discipline. In this work, we assume the SOP has the following characteristics.

- Simplicity and visualization, the worker immediately easily found and understood the necessary instructions.
- OK and not OK (NOK) conditions. An easy way for the operators to visually understand what is OK and NOK when performing their jobs.
- The possibility of rapid changes in process parameters,
- Clarity, which ensures that every worker has all relevant activities in the process as well.
- The ability to monitor the implementation of SOP and its impact on the process parameters.

When all workers at the workstations perform their tasks consistently according to SOPs, less quality, delivery, and safety concerns happen in the production line. In this paper, we discuss how SOP can be used to implement administrative controls in HOC.

2.4 5S

5S [6] stands for five Japanese words: Seiri (sort), Seiton (set in order), Seiso (shine), Seiketsu (standardize), and Shitsuke (sustain), is a system where waste is reduced, safety is improved, and quality are optimized through observing an orderly work area. The 5S defines five steps. The first step (Seiri) sort eliminates unnecessary, broken items from the work area. The second step (Seiton) set in order focuses on providing efficient storage areas for the remaining items. Shine, the third step, is to clean the work area thoroughly. Daily schedules for cleaning the area are created to sustain these changes. Once the first three steps have been implemented, the next step is to standardize the best practices in the work area. In this work, we propose that SOP are created or enforced if already available. The newly developed practices are integrated into the SOP, and they become the standard way of performing actions.

2.5 Problem solving

Manufacturing or safety problems need to be described consistently and systematically to allow for a common understanding and avoid recurrence of same problems. The following steps are a common methodology applied today in industry for problem-solving [8]:

1. Team formation
2. Problem description
3. Interim containment action
4. Root cause analysis
5. Corrective actions
6. Validate corrective actions
7. Implement corrective actions a standardize (update SOP)
8. Team recognition

Problem-solving in our risk assessment model supports continuous improvement and learning loop. During the SOP observations and dialogue between the workers and team leaders, risks and gaps are identified and documented. The problem-solving process is used to learn and continuously improve the way of working at the workstations.

3 Proposed model

The proposed model consists of the the following parts.
a) Risk assessment
b) Risk control strategies
c) Continues improvement and learning with problem solving

The risk assessment model starts with the identification of the hazards. If we take the distance between workers as an example, the hazards would be "social distancing not possible with current workstation set up". In a typical car or truck assembly line, depending on the line-balancing situation and volume requirement, there could be more than two people working closely together, which would be a high-risk situation. To determine and prioritize the risk, a table with the exposure and probability criteria has been defined. See Table 1. The levels of exposure 'E' and probability 'P' should be defined by a cross-functional team of safety experts and operations. Once E and P have been decided, the results which are $E \cdot P$ is an indication of the level of the risk. The higher the value, the higher is the risk. The highest risk should be selected to develop the next steps to define the administration and engineering control actions.

The risk control strategies are based on HOC. While HOC contains five controls, we focus on the engineering controls and administrative control in this paper. Examples of *engineering controls* could be to isolate people from exposure to the virus by, for example, separating people from each other, this can be done by increasing the takt-time (line balancing) and thus need fewer operators at the workstation. Examples of *administrative controls*, can be disinfection of tools before and/or after a work shift, and how operators are expected to move within the workstation to guarantee minimum distance to other operators. Administrative controls can preferably be documented using SOPs to ensure that all operators (experienced and inexperienced) follow the same working standard. It is taken for granted that PPE is already widely in use at all manufacturing sites, which are providing most of the prevention measures today. The control actions, once defined, need to be included as OK and NOK conditions in the SOPs and in the 5S standards to be followed by the team members as they perform the operations at the workstations.

For continuous improvement and learning, SOPs and 5S should be the standard way of working at any manufacturing site to secure that employees adhere to safety and health regulations. Deviations from meeting SOP and requirements can result in errors leading to infection control. The following are some pre-requisites and methods for effective use of the SOPs:

- Clear roles and responsibilities of the supervisors, team leaders, and team members regarding the development, usage, and evaluation of SOPs.
- Current SOP's availability and visibility at locations close to the work areas.
- Clear scheduling for SOP observation. It is recommended daily for team leaders and weekly for supervisors.
- Secure the team members can perform the job according to the SOP by providing dedicated on-the-job training.
- Confirm team members are following the SOP. A job observation between team leaders and the team members of the workstations should be conducted daily for effective SOP compliance.

Table 1. Exposure and probabilities.

Exposure (E)		**Probability** (P)	
Low	1	Improbable	1
Medium	2	Remote	2
High	3	Occasional	3
Very high	4	Probable	4
		Frequent	5

In summary, the proposed framework are as follows:
a) Identification of the hazards.
b) Determine the risk according to the exposure 'E' and the probability 'P', according to Table 1.
c) Determine the level of the risk by multiplying $E \cdot P$.
d) Prioritize the hazards according to the risk level. Follow the sequence below for hazards, starting with those that have the highest level of risk.
e) Determine if any administrative control or engineering control actions are possible.
f) Calculate updated risk levels for the control actions.
g) Define OK and NOK conditions and include in SOP and 5S.
h) Team members at workstations work according to SOP and 5S.
i) Continuously identify deviations from SOP and 5S.
j) Apply problem-solving to improve and resolve deviations.
k) Continue to reduce the level of the hazards.

Next, we will follow up with an example of how the proposed model can be used to evaluate different risk control strategies.

3.1 Example

In Table 2, we can describe HOC via an example, to illustrate more specifically the HOC risk assessment. Lets develop an example with the hazard described in the previous section, which is the social distance not possible with the current workstation setup. Before the type of HOC actions are developed, the risk need to be decided. The cross-functional team decided that this is a high risk of exposure to the virus transmission, which is level 3 ($E = 3$). The team also decided that the level of probability was frequent, which is level 5 ($P = 5$). Therefore the risk level in this example is $E \cdot P = 15$. The first action decided by the team was to reduce the worker density per workstation,

Table 2. Risk assessment model.

Hazards	E	P	$E \cdot P$	Action	new E	new P	new $E \cdot P$
Social distancing not possible with current workstation set up	3	5	15	CA 1: Reduce the workers density per workstation, utilizing the length of the chassis and the width of the cab to meet social distancing requirements wherever possible and max one operator per cab at time. Use face mask all the time and face shields where social distance requirement are not possible.	1	5	5
Social distance not possible with current workstation set up	3	5	15	CA 2: Comply with CA 1 and integrate both the OK and NOK conditions described in SOPs and follow 5S standards.	1	3	3
Social distance not possible with current workstation set up	3	5	15	CA3: Comply with CA1, CA2 and close gaps from SOPs and 5S with problem solving	1	2	2

utilizing the length of the chassis and the width of the cab to meet social distancing requirements wherever possible and max one operator per cab at a time. Use face-mask all the time and face shields where social distance requirements are not possible. The new exposure was reduce to $E = 1$, which is a low exposure and the probability level stayed the same at $P = 5$. Therefore the risk level became after the engineering control action $E \cdot P = 5$, which indicates a reduction of 10 points in the risk level. The next administrative action taken by the team was to use SOPs and 5S to comply with the first action. Examples of SOPs are shown on Fig. 2 and 3. To secure that the people at the workstation use the SOPs, the team decided to use daily SOP and 5S observations. As a result of this new action, the exposure level stayed at level 1, which is low, however, the probability of the risk dropped from frequent to occasional, giving an overall risk improvement from 5 to 3. Finally, during the daily SOP and 5S observations, the teams identified some deviations that could be improved to reduce the virus transmission further. These deviations and gaps were resolved by using problem-solving. After the application of problem-solving the exposure kept a low exposure level 1, however, the probability was further improved to remote, which is level 2. Therefore the new overall risk level became 2.

The previous risk assessment example included a reference to SOP for implementing administrative controls. A similar example could be applied with an SOP for disinfection. Fig. 2 shows the general information about the SOP, while Fig. 3 shows the visual instructions.

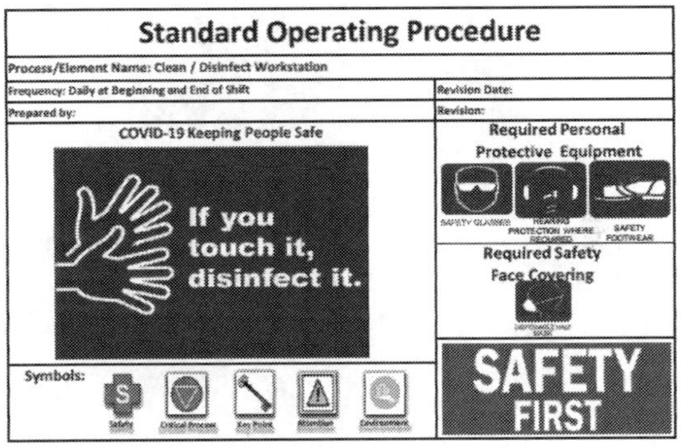

Fig. 2. SOP for disinfection. General information about the SOP.

4 Continuous evaluation

The main objective of the SOP and 5S evaluation is to secure compliance to avoid virus transmission and to secure new SOPs are developed for new identified hazards like, maintenance work would be an issue, job rotation and line balancing. We recommend the following steps to evaluate the SOPs and 5S.

- Capture and document deviations and gaps from the job observations performed between team leaders/supervisors and workers at the workstations. These gaps should be classified according to what is controllable by the teams at the multi-workstations and what is not controllable.
- Visualize and prioritized these gaps so that issues can be resolved using problem-solving.
- Update risk and HOC actions and improve the SOPs based on the gaps that have been found and resolved.
- Visualize the number of gaps open/close, and the number of SOPs updated accordingly. These CI should be correlated to the plant's health and safety

5 Qualitative evaluation

The proposed approach has been evaluated using interviews with line managers at several part distribution centers in the USA. The risk assessment model provides a

https://doi.org/10.1051/matecconf/202032703007

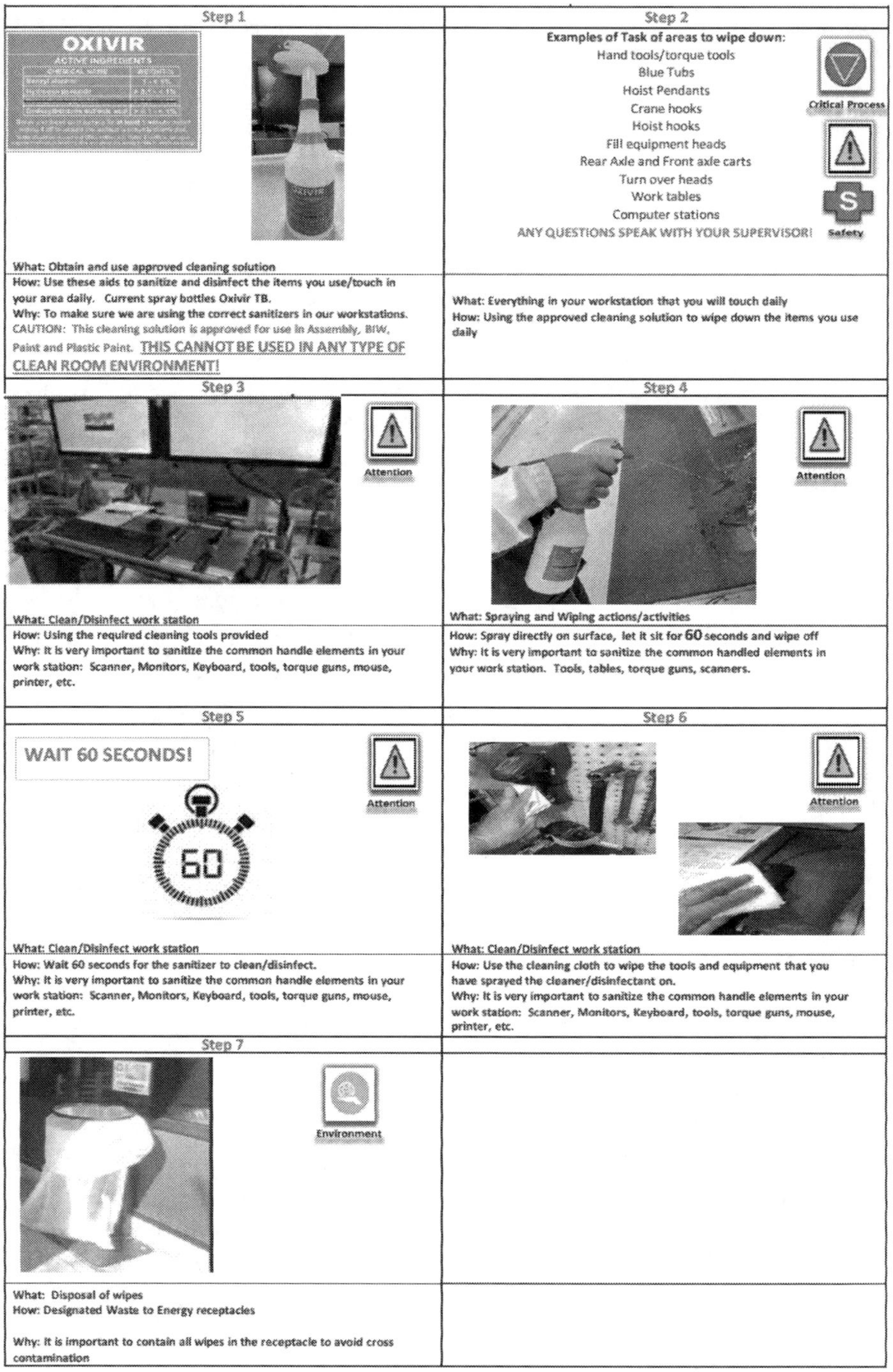

Fig. 3. SOP for disinfection. Detailed instructions for the individual steps.

new way of pandemic prevention, but still at the administrative control level of the HOC pyramid. The SOP focuses on this qualitative evaluation because the SOP precisely described how to perform the job. It does not only helps the operators to understand the sequence of a specific operation. The OK and NOK conditions but also the SOP helps the first line manager to facilitate communication and increasing labor discipline when he conducts the SOP confirmation process.

In our interviews with line managers, we asked the following questions.

- Why do you conduct the SOP confirmation routine, and how COVID-19 is affecting it?
- What are the strength and weaknesses of this process?

We put these questions to several first-line managers in two distribution centers. Here is some feedback from them:

- *Initially, I took for granted that the operators were following the SOP. If we have problems, then I come to discuss with the worker with the help of the SOP if he/she understands the job and if she/he is following the standard. This is the first thing I do to understand a possible problem. During this pandemic, we have been requested to do it at least one time per day, which has helped us a lot to improve the communication with the workers and to secure that the COVID-19 prevention measures are followed.*
- *I have learned a lot with the SOP confirmation process. I was doing it daily because it helped me to train new operators. During COVID-19 we have had a lot of changes of workers. The SOP confirmation process helped me to help the workers. Even when the operators understand the job, we followed the daily routines of SOP confirmation to understand any risk about COVID-19. The workers feel safer when we come and talk to them about COVID-19 risks.*
- *The SOP confirmation routine is helping me to generate ideas for improvement in my area. During COVID-19 we have some problems with the workers not following the prevention measures, especially the use of face mask. During the SOP confirmation process, I was able to identify why some workers did not want to use the face mask.*

The main results of this SOP confirmation process have been the trust of the workers coming to work. Absenteeism has been high with more than 15 percent in some sites, but those sites that have been following the SOP confirmation process the absenteeism has been less than ten percent. As we prepare in the near future to increase production volume, this SOP confirmation will improve both the trust of the workers to be safe, and therefore the absenteeism would be lower.

6 Conclusion

Risk assessment and the corresponding improvement actions should become a highly prioritized process to increase the trust of the workers and employees. This paper presented a methodology based on risk assessment, standard operation procedure, 5S, and problem-solving to help identify and evaluate possible actions to be taken to establish a safe workplace. The model is in the process of being implemented and evaluated in industrial plants. We expect the proposed model to be useful to reduce risk and to capture the new learning like the improvement of the rotation of the workers, increase the versatility of the workers, increase the priority of 5S and SOP at every workstation, increase the frequency of the line balancing process, just to indicate a few examples.

References

1. T. Aven. *Foundations of Risk Analysis*. John Wiley & Sons, 2012.
2. M. Cascella, M. Rajnik, A. Cuomo, S. C. Dulebohn, and R. D. Napoli. Features, evaluation and treatment coronavirus (COVID-19). *StatPearls [Internet]. Treasure Island (FL): StatPearls Publishing; 2020 Jan-*, 2020.
3. Fiat Chrysler Automobiles. Let's keep each other safe – return to work packet. Online; accessed 9-June-2020, 2020.
4. Ford Motor Company. Return to work manufacturing playbook – working together to remain safe and healthy at work. Online; accessed 9-June-2020, 2020.
5. General Motors. Returning to the workplace with confidence. Online; accessed 9-June-2020, 2020.
6. H. Hirano. *5 Pillars of the Visual Workplace: The Sourcebook for 5S Implementation*. New York: Productivity Press, 1996.
7. M. Mĺkva, V. Prajová, B. Yakimovich, A. Korshunov, and I. Tyurin. Standardization – one of the tools of continuous improvement. *Procedia Engineering*, 149:329 – 332, 2016. International Conference on Manufacturing Engineering and Materials, ICMEM 2016, 6-10 June 2016, Nový Smokovec, Slovakia.
8. A. Smalley. *Four Types of problems, from reactive troubleshooting to creative innovation*. Lean Enterprise Institute, Cambridge, MA, USA, 2018.
9. Tesla. Return to work playbook. Online; accessed 9-June-2020, 2020.
10. The National Institute for Occupational Safety and Health (NIOSH). Hierarchy of controls. `https://www.cdc.gov/niosh/topics/hierarchy/default.html`, 2015. [Online; accessed 24-July-2020].

AUTHOR INDEX